Heads Of Lectures On A Course Of Experimental Philosophy

Particularly Including Chemistry

by

Joseph Priestley

ABOUT THE AUTHOR

English scientist, theologian, and political theorist Joseph Priestley (1733–1804) made numerous contributions to chemistry, physics, and philosophy. Priestley, who was raised in Birstall, West Yorkshire, attended local schools before going to Daventry Academy, where he became very interested in natural philosophy. Priestley made significant scientific advances. His greatest-known discovery, made in 1774, was oxygen, which he dubbed "dephlogisticated air." His investigations into gases and their characteristics, which are detailed in "Experiments and Observations on Different Kinds of Air," greatly enhanced our knowledge of chemistry. Priestley also studied nitrous oxide and carbon dioxide, among other gases. In addition to his scientific interests, Priestley was a prolific writer on theological and political topics as well as a dissident clergyman. He received criticism for his Unitarian beliefs and his perceived radical religious ideas. Priestley was a liberal and democratic idealist whose political writings shaped the early American history.

CONTENTS

THE PREFACE

Situated, as I happily am, in the neighbourhood of the *New College at Hackney*, an institution that does honour to the Dissenters, an institution open to all persons without distinction[1], and connected as I am by friendship with the tutors, I was glad to give it every assistance in my power; and therefore undertook to read the *Lectures on History and General Policy* which I had composed when I was tutor at Warrington, and also to give another course on the subject of *Experimental Philosophy*. With this view I drew up the following *Heads of Lectures*; and, to save the students the trouble of transcribing them, they are now printed. To other persons they may serve as a compendious view of the most important discoveries relating to the subject.

As it was found most convenient, with respect to the other business of the college, to confine this course to one lecture in a week, I contrived to bring within that compass as much of the subject of experimental philosophy as I well could, and especially to include the whole of what is called *chemistry*, to which so much attention is now given, and which presents so many new fields of philosophical investigation.

Besides that the plan of the young gentlemen's studies would not admit of it, I think it most advisable not to trouble beginners with more than a large outline of any branch of science. By this means they are not fatigued by too long an attention to any one subject, a greater variety of articles may be brought before them, and in future life they may pursue any of them as much farther as their inclination may dispose, and their ability and opportunity shall enable, them to do it.

I do not give any account of the *experiments* introduced into the several lectures. They will be sufficiently indicated by the subjects of them. They were as many as I could conveniently make within the time; and where the experiments themselves could not be made, I usually exhibited both the different substances employed in them, and those that were the result of them.

As these lectures were calculated for the use of the students at the New College, I prefix an *Address to them*, the same in substance with that which

I delivered to them at the close of the session of 1791. In it may be seen a specimen of the language we hold to them on the subject of *politics*, which with reasonable men will serve as an answer to the many calumnies that have been thrown out against us, as disaffected to the government of this country.

Such institutions will, indeed, always be objects of hatred and dread to *bigots* and the advocates for *arbitrary power*, but the pride of a truly *free country*. I therefore conclude with my earnest prayer (the accomplishment of which the present state of the College does not allow us to doubt) Esto perpetua.

THE DEDICATION

TO THE STUDENTS AT THE NEW COLLEGE IN HACKNEY

My young Friends,

Having drawn up the following *Heads of Lectures* for your use, I take the liberty thus publickly to dedicate them to you; and as I earnestly wish for your improvement and happiness in all respects, excuse me if I take the farther liberty of making a few observations, and giving you some advice, of a more general nature, adapted to your age and circumstances.

As you will soon leave this place of education, and enter upon your several professions and employments, I hope your conduct will demonstrate to the world the solid advantages of this institution, and that the great expence attending it, and the best attention of the managers, have not been bestowed in vain.

Many liberal friends of science, of virtue, and of religion, have contributed to procure you the advantages which you enjoy. They have spared no pains to provide able and careful tutors, and you have had every other advantage for the prosecution of your studies that they could procure you, unclogged by any subscription to articles of faith, or obligation of any other kind, besides such as they have deemed necessary for your own good, and to give the institution its greatest effect. This is an advantage you could not have found elsewhere, at least in this country. And in every seminary of education much more depends upon opportunity for study, free from any obstruction, and undue bias, than upon the ability of tutors; though there is an additional advantage when they are able men, and eminent in the branches of science which they undertake to teach. But this is by no means so essential as many other circumstances.

Whatever be the qualifications of your tutors, your improvement must chiefly depend upon yourselves. They cannot *think* or *labour* for you. They can only put you in the best way of thinking and labouring for yourselves. If, therefore, you get knowledge, you must acquire it by your own industry. You must form all conclusions, and all maxims, for yourselves, from premises and *data* collected, and considered by yourselves. And it is the great object of this institution to remove every bias the mind can lie under, and give the greatest scope to true freedom of thinking and inquiry. And provided you be intelligent and virtuous men, and good citizens, it will be no cause of regret to the friends of this institution, if, with respect to *religion*, or *politics*, you adopt systems of principles, and maxims of conduct, very different from theirs.

Give me leave, now that I am addressing you as *young men*, and young *students*, to suggest a caution on a subject, of the importance of which it is hardly possible that you should be sufficiently aware, because it is only impressed by that *experience* which you have not yet had. I mean that degree of vanity which generally accompanies the acquirements that diligent persons of your age make in places of liberal education, and the contempt they are too apt to entertain for those who have not made the same proficiency with themselves. And I assure you, that in the observations I shall make on this subject, I have no view whatever to any thing that I have observed, or heard, of any of you in particular. But I have been in your situation myself, and I know the importance of these observations to students in general.

You are now at an age in which young persons usually make the most sensible advances in knowledge, and in which the understanding appears to ripen the most rapid manner. You are able to say every year, every month, and almost every day, what particular advances you have made, and how much you know more than you did before. And being taught, and accustomed, to express your thoughts in writing, you find yourselves qualified to do this in a manner of which you had no idea, or expectation, but a little time ago. You also perfectly remember what you have so recently learned, and many respects may be more particular and exact than even your tutors themselves.

The almost unavoidable effect of this is a high idea of your own powers and attainments, and too often a proportionable contempt of those who, not having had equal advantages, cannot do what you are easily capable of. A certain degree of vanity is, therefore, excusable in young persons; and, indeed, it is by means of it that they are excited to exert themselves in a manner that they would not otherwise have done. But be careful that this temper be not indulged to excess, for it will then be found to have serious ill consequences; the least of which is the precluding future improvement, from being already satisfied with yourselves, and conscious of a sufficient superiority over others.

The foundation of this self-conceit, on account of literary attainments, will be found to be extremely weak. In fact, we learn more before the period to which you are now arrived, and I hope you will continue to learn more after it, without its being so much noticed; and the *ability* that is discovered in the acquirements which are the subjects of this vanity is not greater than appears on other occasions. Only they are not so conspicuous.

What we all learn in the first three years of our lives, is much more extraordinary in its nature than all that we acquire afterwards. I mean the perfect use of our limbs, and the elements of speech. What we learn in a month in that early period of life, could not, if we were brought up in the ignorance of it, be learned in a year at any subsequent period. But these acquirements being universal, and what the circumstances in which we are all necessarily in compel us to learn, it does not appear extraordinary in any particular individual. Also, the proficiency that boys make at a grammar school, in which, in general, the dead languages only are taught (a knowledge of which is commonly the result of severe application) is too common to be the cause of much self-conceit. But the advances that are made at places of liberal education are both less common, and of a more conspicuous nature.

You will also find, if you continue your application to study, that it is only the elements of science that you can acquire here, and that if you live many years, they will bear but a small proportion to your future acquirements. But those future acquirements, in consequence of their bearing a less proportion to your whole stock of knowledge, will not be so conspicuous. Thus, though all the buildings that in one year are added to such a city as London would make a pretty large country town, they bear so small a proportion to what was built before, that they are not much noticed; whereas, had half the quantity of building been erected in a place where no house had existed before, it would have been a memorable event in the history of the country.

Also, as in old cities many buildings will fall to decay, while new ones are added; you must expect to forget much of what you now know. No man can give equal attention to every object; and as we advance in life, we, in general, only learn new things at the expence of the old ones. But then they are the more valuable articles of knowledge, the more general and leading principles, which remain with us; while the more useless ones, things to which we give less attention because we find them to be of less use, disappear. Yet it is no uncommon thing for ingenious students to despise old scholars who are not so ready in the *minutiæ* of literature, though they may have forgotten more than those youths ever knew, and may retain what they cannot acquire without forgetting as much.

Another observation proper to lessen the conceit of literary men, is, that genius is not confined to *them*, but is equally great, though not equally conspicuous, in every other line of life, and especially in manufactures and the arts. Here, however, discoveries equal, with respect to *sagacity*, to those of Newton, contribute little to posthumous fame, because the discoverers are not writers.

But the greatest branch of intellectual excellence with respect to which every other is nothing, and which, from its nature, can never be foundation of any self-conceit, is *virtue*, or right dispositions of mind, leading to right conduct in life. Proper sentiments, and just affections of mind, arise from just, and often comprehensive, views of things, past, present, and to come. And if the real greatness of any thought, or action, be estimated by the number of elements that constitute it, and its remoteness from the dictates of sense and natural appetite, a virtuous and pious man will appear to be a much more dignified character, more proper to be viewed with admiration and esteem, than the greatest scholar; discovering, in fact, greater comprehension and force of mind. I mean, however, that virtue which is the result of reflection, of discipline, and much voluntary exertion, which, though operating with equal promptness and facility, is as much superior to mere *innocence*, and what is commonly called *good nature*, as motions secondarily automative are to those that are primarily so; a comparison which you who have studied *Hartley's Theory of the Mind* will see the force of.

These considerations I take the liberty to suggest, as being proper to lessen that vanity which is so incident to those who distinguish themselves in the fields of literature, and which, operating like the acquisition of riches, or power, or any possession that is *rare* among men, instead of enlarging, may tend to contract the mind, by confining its attention to itself. Beginning with a generous emulation, it proceeds to envy and jealousy, and ends in actual hatred and malignity, against which you cannot, surely, be too strongly put upon your guard; this being the greatest depravity to which human nature is subject, and which yet, like any other vice, may be in full possession of the mind, without the person himself knowing, or suspecting it; unless he give more attention to his feelings than most persons do. If no man ever thought himself to be avaricious, or cruel, can it be expected that any person should ever discover that he is too self-conceited?

Better, however, infinitely better, were it to rank with the merest dunces, than have the conceit and malignity (produced originally from conceit) of some who have distinguished themselves the most as linguists, critics, and poets. Even the study of nature, though, from its vast extent, it is less apt to produce this baneful effect, is not always a sufficient guard against it. This is an affecting and an alarming consideration. But in the intellectual, as in the civil and commercial world, we gain nothing but at the risk of some loss; and in this case the possible gain is worth the risk of even this great loss.

For when literary, and scientific excellence coincide with that which is of a moral nature, it adds unspeakably to the value of a character. Ingenuity coupled with modesty, and great genius with benevolence and true piety, constitute the perfection of human character, and is what we should ever have in view. And a course of education in which both these objects are equally attended to, is the only one that deserves to be called *liberal*: but such as, I hope, you have found this to be.

Give me leave further to observe to you, that the time that you spend in a place of liberal education is of more importance to you than you can be at present aware of. Whatever be the sphere of life for which you are destined, the probability is, that you will hereafter have but little leisure for reading and studying, compared to what you have now. Besides, general maxims of all kinds, such as are the foundation of all our future conduct, in morals, religion, or politics, are generally formed at your time of life. From this period expect no great change in your opinions, or conduct; because now it is that you give particular attention to the forming your opinions on all subjects of importance; so that very little that is materially new to you can be expected to occur to you in future life, and almost every thing that you would choose to read will only tend to confirm you in the general principles that you will now adopt. There are, no doubt, exceptions to this, as well as every other general observation; but it is wisdom to suppose, and to act upon the supposition, that we are constituted as the generality of mankind are, and that we shall feel, and act, as they do. Since, then, so much depends on the leading principles and maxims which you will now adopt, be it your care to form just and good ones, and let no authority of tutors, or others, have any undue influence over you. In all cases think and judge for yourselves, and especially on all subjects of importance, and with as much attention as you can give to them.

It may not be amiss, in the present state of things, to say something respecting another subject, which now commands universal attention. You cannot but be apprised, that many persons entertain a prejudice against this College, on account of the republican, and, as they choose to call them, the licentious, principles of government, which are supposed to be taught here. Show, then, by your general conversation, and conduct, that you are the friends of peace and good order; and that, whatever may be your opinions with respect to the best form of government for people who have no previous prejudices or habits, you will do every thing in your power for the preservation of that form of it which the generality of your countrymen approve, and under which you live, which is all that can be reasonably expected of any subject. As it is not necessary that every good son should think his parent the wisest and best man in the world, but it is thought sufficient if the son pay due respect and obedience to his parent; so neither is it to be expected that every man should be of opinion that the form of government under which he happens to be born is the best of all possible forms of government. It is enough that he submit to it, and that he make no attempt to bring about any change, except by fair reasoning, and endeavouring to convince his countrymen, that it is in their power to better their condition in that respect, as well as in any other. Think, therefore, speak, and write, with the greatest freedom on the subject of government, particular or general, as well as on any other that may come before you. It can only be avowed tyranny that would prevent this. But at the same time submit yourselves, and promote submission in others, to that form of government which you find to be most approved, in this country, which at present unquestionably is that by *King, Lords, and Commons.*

As to *religion*, we may, surely, be allowed to think and act entirely for ourselves; in all cases obeying God and conscience rather than man. But let us be thankful for the degree of liberty that we are allowed, though it be not all that we are justly entitled to; and let us not use any other means than reason and argument in order to better our condition. By this peaceable and steady conduct we shall at length convince those who will hear reason, that we are entitled to greater consideration; and doubt not but whatever is *true* and *right*, will finally prevail, and be universally established.

That any of your tutors, or any of the friends of this institution, wish to promote reformation, in church or state, by any other means than those of reason, and argument, is a *calumny*, utterly void of foundation, or probability. But your conduct, dispersed as you will soon be in different parts of the country, will be the best means of refuting it. Let us leave the method of proceeding by *riot* and *tumult* to those persons to whose schemes such proceedings are congenial. Truth stands in no need of such support, and will always triumph when assailed by such weapons. In return, then, for the advantages which you have enjoyed in this institution, do it this service; and in recommending it, I trust you are doing substantial service to the cause of liberty and truth; and conferring a most important benefit on your country, and on mankind.

LECTURE I

The Introduction

The object of experimental philosophy is the knowledge of nature in general, or more strictly, that of the properties of natural substances, and of the changes of those properties in different circumstances. This knowledge can only be attained by *experiment,* or *observation*; as that clay is capable of becoming hard by means of fire, and thereby being made into bricks, and that by the same means lime-stone can be converted into quick-lime, and by the addition of water and sand, make mortar. It is by observation also that we discover that stones and other heavy bodies fall to the ground, and that a magnet will attract iron. In other words, experimental philosophy is an investigation of the wisdom of God in the works and *the laws of nature,* so that it is one of the greatest objects to the mind of man, and opens a field of inquiry which has no bounds; every advance we make suggesting new doubts and subjects of farther inquiry.

The uniformity we discover in the properties of natural substances enables us to lay down general rules, or principles, which, being invariable, we call the laws of nature; and by our knowledge of these laws we are able to predict, and at our own pleasure to produce, particular results, and this is the source of all the powers of man. It is the direction we acquire of the powers of nature; so that, as Lord Bacon observed, *knowledge is power.*

All arts and manufactures are derived from science. Thus the doctrine of *mechanics* is an application of the law of gravitation. Every thing we are capable of doing by means of the steam-engine is derived from our knowledge of the properties of water in steam; all the great effects of gunpowder we owe to our knowledge of the composition, and chemical properties, of that substance.

Every new appearance in nature is preceded by some new circumstance, and to this, or rather to something always attending it, we say that the appearance is *owing*. This circumstance we therefore call the *cause,* and the new appearance the *effect* of that cause. Thus we say that the union of phlogiston to a particular kind of earth is the cause of its becoming a metal.

It is one of the principal rules of philosophizing to admit no more causes than are necessary to account for the effects. Thus, if the power of gravity, by which heavy bodies fall to the earth, be sufficient to retain the planets in their orbits, we are authorized to reject the *Cartesian Vortices*. In other words, we must make no more general propositions than are necessary to comprehend all the particulars contained in them. Thus, after having observed that iron consists of a particular kind of earth united to phlogiston, and that it is soluble in acids; and that the same is true of all other metallic substances, we say, universally, that all metals consist of a peculiar earth and phlogiston, and that they are all soluble in some acid.

Of the circumstances which occasion a change in the properties of bodies, some are the addition of what are properly called *substances*, or things that are the objects of our senses, being *visible*, *tangible*, or having *weight*, &c. Thus the addition of an acid changes an alkali into a neutral salt. But other changes are occasioned either by a change of texture in the substance itself, or the addition of something that is not the object of any of our senses. Thus, a piece of steel becomes a magnet by the touch of another magnet, and a drop of glass acquires the property of flying asunder by a small fracture, in consequence of falling when red hot into cold water. Such also, in the opinion of some, is the difference between hot and cold substances.

Till the nature of the cause be ascertained, it is convenient to make use of the term *principle*, as including both of the above-mentioned causes of the change of properties in bodies. Thus, whatever be the real cause of gravity, or of inflammability, we may speak of the *principle of gravity*, or of *inflammability*; whether, with Newton, we suppose gravity to be occasioned by a fluid pervading the whole universe, which he termed æther, and whether inflammability be caused by the presence of a real substance called phlogiston, or not. In this manner we use the letters x and y to denote unknown quantities in algebra.

When changes are made in substances by the addition of other substances, they make what is called a *chemical union*; and in this case the properties of the compound cannot with any certainty be deduced from those of the component parts, but must be ascertained by fresh experiments. Thus, from the specific gravities, or the degrees of fusibility, of two metals, those of the compound cannot be predicted. Neither water nor acid of vitriol will separately dissolve iron, so as to produce inflammable air, but both together will do it. However, the properties of similar compounds are similar to one another. Thus, all metals dissolved in acids are precipitated by mild alkalis. This chemical union of two substances we ascribe to a certain *elective attraction*, or *affinity* that subsists between them, in consequence of which they unite with one another whenever a proper opportunity offers, in preference to those substances to which they were before united. Thus the vitriolic acid, having a stronger affinity with the vegetable alkali which is the basis of nitre, will unite with that alkali, and with it form another compound, called *vitriolated tartar*, while the acid of nitre, being detached from its base, is collected separately.

When two substances compose one liquid, and a third, which has a stronger affinity with either of the two parts than they have with each other, is added to them, it will unite with that part, and take its place in the solution, while the other will in many cases be precipitated, and may be collected. Thus the earth of alum is precipitated from a solution of alum by salt of tartar. This is the case of *simple affinity*.

When both the substances are compounds, the component parts of which have a weaker affinity with each other than they have with those of the other compound, two new combinations are formed, and this is called a case of *double affinity*. Thus when phlogisticated alkali is poured into a solution of green vitriol, the acid of the vitriol unites with the alkali, while the phlogiston joining the calx of iron makes Prussian blue.

All nature lying open to our investigation, we must consider the different parts in some order. But it is not very material which we adopt, because, begin where we will, the properties of the substances we first treat of will be connected with those which must be particularly considered afterwards, the changes in one substance being occasioned by its union with another. It will be impossible, for example, to explain the properties of metals without considering the *acids*, because by their union with acids very important changes are made in their properties.

There have been three principal methods of arranging natural substances. One is according to the *three kingdoms*, as they are called, into which they have been distributed, viz. the *mineral*, the *vegetable*, and the *animal*. Another is according to the *elements* which enter into their composition, and a third according to the *form* in which they are usually found, viz. *aerial, fluid*, or *solid*. Upon the whole this last appears to me to be the most convenient, especially as it is easy to intermix general observations concerning the other divisions when they are particularly wanted. As there will be frequent occasion to speak of the component and elementary parts of all substances, I shall here observe, that, according to the latest observations, the following appear to be the elements which compose all natural substances, viz. *dephlogisticated air*, or the *acidifying principle*; *phlogiston*, or the *alkaline principle*; the different *earths*, and the principles of *heat, light,* and *electricity*. Besides these, there are the following principles which have not been proved to be substances, viz. *attraction, repulsion,* and *magnetism*. By the help of these principles we are able, according to the present state of natural knowledge, to explain all the appearances that have yet occurred to us.

LECTURE II

Of the Properties of all Matter

Before I consider the properties of particular substances, it will be proper to mention those which are common to them all. But I shall first observe, that the term *substance* has no proper idea annexed to it, but is merely a convenience in speech; since we cannot speak of the properties of substances, such as *hard, round, coloured*, &c. &c. (which circumstances alone affect our senses, and thereby give proper ideas) without saying that they inhere in, or belong to, some *thing, substance*, or *substratum*. The terms *being* and *person* are also in the same predicament.

One property of all substances is *extension*, since they all occupy some portion of space.

The incapacity of any substance to change its place has been termed, though improperly, the *vis inertiæ* of matter. It is sufficient to say, that neither this, nor any other effect can be produced without a cause.

Infinite divisibility is a necessary property of all extended substance; and from this circumstance it will follow, that the smallest quantity of solid matter may be made to fill the largest space, and yet none of the pores shall exceed the smallest given magnitude; and consequently, that, for any thing we know to the contrary, all the bodies in the universe may be comprized in the smallest space.

Another property usually ascribed to all matter is *impenetrability*, or the necessary exclusion of any substance from the place occupied by another. But the only proof of impenetrability is the *resistance* that we find to our endeavours to put one substance into the place of another; and it is demonstrated by experiments, that this resistance is not occasioned by the actual contact of the substances, but by a power of repulsion acting at a real distance from their surfaces. It requires a considerable force to bring two solid substances into as near contact as the particles of the same substance; and that *these* are not in actual contact is evident, from their being capable of being brought nearer by cold; and this is most remarkable with respect to the heaviest, that is, the densest, of all substances, viz. the metals.

A more positive argument for the penetrability of matter is, that the particles of light, after entering the densest transparent substance, do not appear to meet with any obstruction to their progress till they come to the opposite side.

The powers of *attraction* and *repulsion* seem to be common to all matter, and the component parts of all substances are kept in their places by the due balance of those opposite powers. If, by any means, the particles of any substance be removed beyond their sphere of mutual attraction, they repel one another, as those of water when it becomes steam.

Of the different kinds of attraction, that of *gravitation* seems to extend to the greatest possible distance; but that which keeps together the parts of the same substance, thence called the *attraction of cohesion*, and the different kinds of chemical attractions, called *affinities*, only act at a small distance. Of the causes of these attractions we are entirely ignorant.

Of Aeriform Substances

Aeriform substances, of which the air that we breathe is one, though invisible, are real substances, as appears by their excluding other substances.

That the air has *weight* appears by actually weighing a vessel before and after it is exhausted of air by means of an air-pump (an instrument contrived for that purpose) by its bursting a bladder, and various other experiments.

Air, being a fluid, presses in all directions, as in the experiment of the fountain in *vacuo*, and others.

The weight of the air is the cause of the suspension of mercury in a barometer, and of the action of pumps. The weight of atmospherical air is to that of water in the proportion of about 1 to 800, so as to press with the weight of about fourteen pounds on every square inch of surface.

Air, being an elastic fluid, is capable of occupying more or less space according to the pressure which it sustains, as appears by a bladder partially filled with air being expanded when the air is drawn from a receiver in which it is put, by means of the air-pump, and compressed in the condensing engine, an instrument the reverse of the air-pump.

Air is necessary to the conveyance of sound, to the existence of flame, and to animal life.

LECTURE III

Of Atmospherical Air

The first species of air that offers itself to our consideration is that of the atmosphere, which appears to consist of a mixture of two kinds of air, of different and opposite qualities, viz, dephlogisticated and phlogisticated, in the proportion of about one third of the former to two thirds of the latter. It is by means of the former of these two ingredients that it is capable of supporting flame and animal life.

This composition of atmospherical air is proved by several substances absorbing the dephlogisticated air, and leaving the phlogisticated. All these processes have been termed *phlogistic*, because the effect is not produced but by substances supposed to contain phlogiston in a volatile state; and by the affinity between phlogiston and the dephlogisticated part of the air, the one is separated from the other. Of these processes are the calcination of metals, a mixture of iron-filings and sulphur, liver of sulphur, the burning of phosphorus, and the effluvia of flowers.

In some cases, however, it is not so clear that any thing is emitted from the substance that produces this effect; for water deprived of all air will absorb the dephlogisticated part of the atmospherical in preference to the phlogisticated part.

As the purity of atmospherical air, or its fitness for respiration, depends upon the proportion of the dephlogisticated air that it contains, any of the above-mentioned processes will suffice to determine it. The more any given quantity of air is diminished by any of them, the purer it was before the diminution. But this effect is produced the most quickly by a mixture of nitrous air, or firing inflammable air in it, being almost instantaneous.

In order to measure the purity of air, it is convenient to take more of the nitrous or inflammable air than is necessary to saturate the dephlogisticated air it contains. Equal quantities of each best answer the purpose. Supposing a given quantity of atmospherical air to be mixed with an equal quantity of nitrous air, and the residuum to be 1.1 measure, the proportion of dephlogisticated and phlogisticated air in it may be found by the following arithmetical operation, it being here taken for granted that one measure of pure dephlogisticated air will saturate two measures of pure nitrous air.

> 2.0 viz. one of each.
> 1.1 the residuum.
>
> — —
>
> 3)0.9 the quantity that has disappeared.
> 0.3 the dephlogisticated air contained
> in the measure of the air examined.

And this substracted from 1 leaves .7 for the proportion of phlogisticated air in it.

LECTURE IV

Of Dephlogisticated Air

Dephlogisticated air, which is one of the component parts of atmospherical air, is a principal element in the composition of acids, and may be extracted by means of heat from many substances which contain them, especially the nitrous and vitriolic; as from nitre, red precipitate, the vitriols, and turbith mineral, and also from these two acids themselves, exposed to a red heat in an earthen tube. This kind of air is also contained in several substances which had attracted it from the atmosphere, as from precipitate *per se, minium, & manganese.*

Dephlogisticated air is likewise produced by the action of light upon green vegetables; and this seems to be the chief means employed by nature to preserve the purity of the atmosphere.

It is this ingredient in atmospheric air that enables it to support combustion and animal life. By means of it the most intense heat may be produced, and in the purest of it animals will live nearly five times as long as in an equal quantity of atmospherical air.

In respiration, part of this air, passing the membrane of the lungs, unites with the blood, and imparts to it its florid colour, while the remainder, uniting with phlogiston exhaled from the venous blood, forms fixed air. It is dephlogisticated air combined with water that enables fishes to live in it.

Dephlogisticated air is something heavier than atmospherical air, and the purity of it measured by mixing with it two equal quantities of nitrous or inflammable air, deducing the residuum after the diminution from the three measures employed, and dividing the remainder by 3, as in the process for common air.

Of Phlogisticated Air

The other ingredient in the composition of atmospherical air is phlogisticated air. It is procured by extracting the dephlogisticated part of the common air, as by the calcination of metals, &c. &c. by dissolving animal substances in nitrous acid, and also by the union of phlogiston with nitrous air, as by heating iron in it, and by a mixture of iron-filings and sulphur.

Phlogisticated air extinguishes a candle, is entirely unfit for respiration, and is something lighter than common air. It is not capable of decomposition, except by exploding it together with a superabundance of dephlogisticated air, and a quantity of inflammable air, or by taking the electric spark repeatedly in a mixture of it and dephlogisticated air. In these cases nitrous acid is formed.

LECTURE V

Of Inflammable Air

Inflammable air is procured from all combustible substances by means of heat and water, and from several of the metals, especially iron, zink, and tin, by the vitriolic and marine acids.

From oils and spirit of wine it is procured by the electric spark. By the same means also alkaline air is converted into it.

That which is procured from metals, especially by steam, is the purest and the lightest, about ten times lighter than common air; in consequence of which, if a sufficient quantity be confined in a light covering, it is possible to make it carry up heavy weights.

When it is procured from animal or vegetable substances, it is of a heavier kind, and burns with a lambent flame, of various colours, according to the circumstances.

Calces of metals heated in inflammable air are revived, and the air absorbed; and since all the metals are revived in the same inflammable air, the principle of metallization, or *phlogiston*, appears to be the same in them all.

Pure inflammable air seems to consist of phlogiston and water, and the lambent kinds to be the same thing, with the addition of some oily vapour diffused through it.

LECTURE VI

Of Nitrous Air

Nitrous air is procured by dissolving most of the metals, especially iron, mercury, and copper, in the nitrous acid; but that from mercury seems to be the purest. Nitrous air produced from copper contains a mixture of phlogisticated air. Some nitrous air is also obtained from the solution of all vegetable substances in nitrous acid; whereas animal substances in the same process yield chiefly phlogisticated air: but in both these cases there is a mixture of fixed air.

This species of air is likewise produced by impregnating water with nitrous vapour. This process continues to have this effect after the water becomes blue, but ceases when it turns green; there not then, probably, being a sufficient proportion of water. Nitrous air is likewise produced by volatile alkali passing over red hot manganese, or green vitriol, when they are yielding dephlogisticated air. This shews that dephlogisticated air is one ingredient in the composition of nitrous air, and the same thing appears by pyrophorus burning in it. On the contrary, when nitrous air is made to pass over red-hot iron, volatile alkali is produced.

Nitrous air is completely decomposed by a mixture of about half its bulk of dephlogisticated air, and the produce is nitrous acid. And as nitrous acid is likewise formed by the union of inflammable and dephlogisticated air, one principal ingredient in nitrous air must be common to it and inflammable air, or phlogiston. This air is likewise decomposed by dephlogisticated nitrous acid, which by this means becomes phlogisticated. It is also decomposed by a solution of green vitriol, which by this means becomes black, and when exposed to the air, or heated, emits nitrous air, and recovers its former colour. These decompositions of nitrous air seem to be effected by depriving it of phlogiston, and thereby reducing it to the phlogisticated air originally contained in it.

This kind of air is diminished to about one fourth of its bulk by a mixture of iron filings and brimstone, or by heating iron in it, or calcining other metals in it, when the remainder is phlogisticated air. All that iron gets in this process is an addition of weight, which appears to be water, but it loses its phlogiston, so that nitrous air seems to contain more phlogiston, and less water than phlogisticated air.

Nitrous air and dephlogisticated air will act upon one another through a bladder, but in this case there remains about one-fourth of the bulk of nitrous air, and that is phlogisticated air; so that in this case there seems to be a conversion of nitrous air into phlogisticated air without any addition of phlogiston.

Nitrous air is decomposed by pyrophorus, and by agitation in olive oil, which becomes coagulated by the process. It is also absorbed by spirit of turpentine, by ether, by spirit of wine, and alkaline liquors.

It is imbibed by charcoal, and both that air which is afterwards expelled from it by heat, and that which remains unabsorbed, is phlogisticated air.

Nitrous air resists putrefaction, but is diminished by the animal substances exposed to it to about a fourth of its bulk, and becomes phlogisticated air. It is likewise fatal to plants, and particularly to insects.

When nitrous air is long exposed to iron, it is diminished and brought into a state in which a candle will burn in it, though no animal can breathe it. But this peculiar modification of nitrous air, called *dephlogisticated nitrous air*, is produced with the greatest certainty by dissolving iron in spirit of nitre saturated with copper, impregnating water with this air, and then expelling it from the water by heat. If bits of earthen ware be heated in this dephlogisticated nitrous air, a great proportion of it becomes permanent air, not miscible with water, and nearly as pure as common air, so that the principle of *heat* seems to be wanting to constitute it permanent air.

LECTURE VII

Of Fixed Air

Having considered the properties of those kinds of air which are not readily absorbed by water, and therefore may be confined by it, I proceed to those which *are* absorbed by it, and which require to be confined by mercury. There are two kinds, however, in a middle state between these, being absorbed by water, but not very readily; a considerable time, or agitation, being necessary for that purpose. The first of these is *fixed air*.

This kind of air is obtained in the purest state by dissolving marble, lime-stone, and other kinds of mild calcareous earth in any acid. It is also obtained by the burning, or the putrefaction, of both animal and vegetable substances, but with a mixture of both phlogisticated and inflammable air. Fixed air is also produced by heating together iron filings and red precipitate; the former of which would alone yield inflammable air, and the latter dephlogisticated. Fixed air is therefore a combination of these two kinds of air.

Another fact which proves the same thing is, that if charcoal of copper be heated in dephlogisticated air, almost the whole of it will be converted into fixed air. On the same principle fixed air is produced when iron, and other inflammable substances, are burned in dephlogisticated air, and also when minium, and other substances containing dephlogisticated air, are heated in inflammable air.

That water is an essential part of fixed air is proved by an experiment upon *terra ponderosa aerata*, which yields fixed air when it is dissolved in an acid, but not by mere heat. If steam, however, be admitted to it in that state, it will yield as much fixed air as when it is dissolved in an acid.

Water absorbs something more than its own bulk of fixed air, and then becomes a proper acid. Iron dissolved in this water makes it a proper chalybeate; as without iron it is of the same nature with Pyrmont or Seltzer water, which by this means may be made artificially.

Ice will not imbibe this air, and therefore freezing expels it from water.

Fixed air extinguishes flame, and is fatal to animals breathing in it. Also water impregnated with this air is fatal to fishes, and highly injurious to plants. But water thus impregnated will prevent, in a great measure, the putrefaction of animal substances.

Fixed air thrown into the intestines, by way of glyster, has been found to give relief in some cases of putrid disease.

Of Hepatic Air

Another species of air absorbed by water, but not instantly, is termed *hepatic air*, being produced by the solution of liver of sulphur, or of sulphurated iron, in vitriolic or marine acid.

Water imbibes about twice its bulk of this kind of air, and it is then the same thing with the sulphureous waters of Harrowgate.

Of Phosphoric Air

Phosphoric air is produced by the solution of phosphorus in caustic fixed alkali. If this air be confined by mercury, it will take fire on being admitted to atmospheric, and much more to dephlogisticated air. After agitation in water it loses this property, and the residuum is merely inflammable air, with no great diminution of its bulk. This kind of air, therefore, probably consists of phosphorus dissolved in inflammable air; though it cannot be made by melting it in inflammable air.

LECTURE VIII

Of Dephlogisticated Marine Acid Air

This species of air is produced by heating spirit of salt with manganese; or more readily, by pouring acid of vitriol on a mixture of salt and manganese, in the proportion of about 16 of the former to 6 of the latter. In this case the acid of vitriol decomposes the salt, and the marine acid, disengaged in the form of air, takes dephlogisticated air from the manganese; so that this species of air seems to consist of marine acid vapour, and dephlogisticated air.

This species of air has a peculiarly pungent smell, and is absorbed by water as readily as fixed air.

The water takes about twice its bulk of it; and thereby acquires a yellowish tinge. Both this air, and the water impregnated with it, discharges vegetable colours from linen or cotton, and is thereby useful in bleaching.

This air when cold coagulates into a yellowish substance. It dissolves mercury, and with it forms *corrosive sublimate*.

Of Phlogisticated Marine Acid Air

Besides the preceding kinds of air which are slowly absorbed by water, there are others which are absorbed by it very rapidly, so that they cannot be confined but by mercury.

Of this kind is *phlogisticated marine acid air*, procured by the acid of vitriol and common salt; the former seizing upon the alkaline basis of the latter, and thereby expelling the marine acid in the form of air.

It is called *phlogisticated* to distinguish it from *dephlogisticated marine acid air*, which seems to be the same thing, with the addition of dephlogisticated air.

Phlogisticated marine acid air is heavier than common air. It extinguishes a candle with a blue flame. It dissolves many substances containing phlogiston, as iron, dry flesh, &c. and thereby forms a little inflammable air. Water absorbs 360 times its bulk of this air, and is then the strongest spirit of salt. It absorbs one-sixth more than its bulk of alkaline air, and with it forms the common sal ammoniac. Its affinity to water enables it to dissolve ice, and to deprive borax, nitre, and other saline substances, of the water that enters into their composition.

LECTURE IX

Of Vitriolic Acid Air

Vitriolic acid air is procured by heating in hot acid of vitriol almost any substance containing phlogiston, especially the metals which are soluble in that acid, as copper, mercury, &c. This kind of air is heavier than common air, and extinguishes a candle, but without any particular colour of its flame. It will not dislodge the nitrous or marine acids from any substance containing them.

By its affinity to water it deprives borax of it.

One measure of this air saturates two of alkaline air, and with it forms the vitriolic ammoniac.

Water imbibes between 30 and 40 times its bulk of this air, and retains it when frozen. Water thus impregnated dissolves some metals, and thereby yields inflammable air.

If this water be confined in a glass tube, together with common air, and be exposed to a long continued heat, it forms real sulphur, the dephlogisticated part of the common air being imbibed, and forming real vitriolic acid, which uniting with the phlogiston in the air, forms the sulphur. Also this air mixed with atmospheric air will, without heat, imbibe some part of it, and thereby become the common acid of vitriol; so that water impregnated with vitriolic acid air, commonly called *sulphureous,* or *phlogisticated acid of vitriol,* wants dephlogisticated air to make it the common acid of of vitriol.

This kind of air is imbibed by oils, which thereby change their colour; whale oil becoming red, olive oil of an orange colour, and spirit of turpentine of the colour of amber.

If this air be confined in a glass tube by mercury, and the electric spark be taken in it, a black tinge will be given to the glass contiguous to the spark, and this black substance appears to be mercury super-phlogisticated; since by exposure to air it becomes running mercury: so that the vapour of mercury must be diffused through every part of this air, to the distance of at least several feet from the surface of the mercury.

Of Fluor Acid Air

Fluor acid air is procured by dissolving the earthy substance called *fluor* in vitriolic acid.

This kind of air extinguishes a candle, and, like vitriolic acid air, one measure of it saturates two of alkaline air. It is peculiar to this kind of air to dissolve glass when it is hot.

It seems to consist of a peculiar acid vapour united to the stony substance of the fluor; for water being admitted to it absorbs the acid vapour, and the stony substance is deposited. By this means it exhibits an amusing appearance, whether water be admitted to a glass jar previously filled with that air, or the bubbles of air be admitted, as they are formed, to a quantity of water resting on mercury.

LECTURE X

Of Alkaline Air

Alkaline air is produced by means of heat from caustic volatile alkali, and also from a mixture of sal-ammoniac and slaked lime, in the proportion of about one-fourth of the former to three-fourths of the latter. In this case the marine acid in the sal-ammoniac unites with the calcareous earth, and the volatile alkali (probably with the assistance of the water) takes the form of air.

This species of air is heavier than inflammable air, but lighter than any of the acid airs. Like them, however, it dissolves ice, and deprives alum, and some other saline substances, of the water which they contain. United with fixed air, it makes the concrete volatile alkali; with marine acid air, the common sal-ammoniac; and with water, the caustic volatile alkali.

The electric spark, or a red heat, converts alkaline air into three times its bulk of inflammable air; and the calces of metals are revived in alkaline, as well as in inflammable air; but there remains about one-fourth of its bulk of phlogisticated air. These facts shew that alkaline air consists chiefly of phlogiston.

Miscellaneous Observations relating to Air

The *nitrous* acid may be exhibited in the form of air, as well as the vitriolic, the marine, and the fluor acids. But it cannot be confined even by mercury, which it instantly dissolves. It may, however, in some measure, be confined in a dry glass vessel, from which it will in a great measure expel the common air. This nitrous acid air is that red vapour, which is produced by the rapid solution of bismuth, and some other metals in the nitrous acid. But the vegetable acid cannot be exhibited in the form of air. It is only capable of being converted into vapour, like water: and in the common temperature of our atmosphere, returns to a state of fluidity.

Different kinds of air which have no affinity to each other, when once mixed together will not separate, notwithstanding any difference of specific gravity. Such is the case of a mixture of inflammable and dephlogisticated air, and even of inflammable and fixed air. Without this property also, the phlogisticated air, which constitutes the greatest part of our atmosphere, being specifically lighter than dephlogisticated air, of which the other part of it consists, would separate from it, and ascend into the higher regions of the atmosphere. Inflammable air, however, will not mix with acid or alkaline air.

Different kinds of air are expanded differently by the same degrees of heat; dephlogisticated air the least, and alkaline air the most.

If any fluid, as water, spirit of wine, or even mercury, be heated in a porous earthen vessel, surrounded by any kind of air, the vapour of the fluid will pass through the vessel *one* way, while the air passes the *other*; and when the operation ceases, with respect to the *one*, it likewise ceases with respect to the *other*.

LECTURE XI

Of Liquid Substances;
AND FIRST OF
WATER

Having considered all the substances that are usually found in the form of *air*, I come to those that are generally in a *fluid* form, beginning with *water*, which is the principal, if not the only cause of fluidity to all the other substances that I shall place in this class.

Pure water is a liquid substance, transparent, without colour, taste, or smell; and with different degrees of heat and cold may be made to assume the three forms of a solid, of a fluid, and of air. Below 32° of Fahrenheit it is ice, and above 212° it is vapour; so that in an atmosphere below 32° it never could have been known to be any thing else than a peculiar kind of stone, and above 212° a peculiar species of air.

In passing from the state of a solid to that of a liquid, water absorbs a great quantity of the principle, or matter, of *heat*, which remains in it in a *latent* state; and in passing from a state of fluid to that of vapour, it absorbs much more; and this heat is found when the processes are reversed. It has been observed, that when water becomes vapour, it takes the form of small globules, hollow within, so as to be specifically lighter than air.

The degree of heat at which water is converted into vapour depends upon the pressure of the atmosphere; so that in vacuo, or on the top of a high mountain, it boils with little heat; and when compressed, as in Papin's digester, or in the bottom of a deep pit, it requires much heat. In the former case the restoring of the pressure will instantly put a stop to the boiling, and in the latter case the removing of the pressure will instantly convert the heated water into vapour.

The ease with which water is converted into vapour by heat, has given a great power to mechanicians, either by employing the natural pressure of the atmosphere, when steam is condensed under a moveable pistern, in an iron cylinder, which was the principle of the old fire-engine, or by employing the elastic power of steam to produce the same effect, which is the principle of Mr. Watt's steam engine.

Water was long thought to be incompressible by any external force, but Mr. Canton has shewn that even the pressure of the atmosphere will condense it very sensibly.

We do not know any external force equal to that by which water is expanded when it is converted into ice, or into vapour. For though the particles of water approach nearer by cold, yet when it crystallizes, the particles arrange themselves in a particular manner, with interstices between them; so that, on the whole, it takes up more room than before.

Water has an affinity to, and combines with, almost all natural substances, aerial, fluid, or solid; but most intimately with acids, alkalies, calcareous earth, and that calx of iron which is called *finery cinder*, from which the strongest heat will not expel it.

It has been supposed by some, that by frequent distillation, and also by agitation, water may be converted into a kind of earth; but this does not appear to be the case. It has also of late been thought, that water is resolvable into dephlogisticated and inflammable air; but the experiments which have been alleged to prove this do not satisfy me; so that, for any thing that appeared till very lately, water might be considered as a simple element. By means of heat, however, it seems to be resolvable into such air as that of which the atmosphere consists, viz. dephlogisticated and phlogisticated, only with a greater proportion of the former.

Water, with respect to specific gravity and temperature, has generally been made the standard to all other substances; its freezing and boiling points being the limits by means of which thermometers are graduated. Other substances have also been compared with water, as a standard, with respect to the capacity of receiving heat, and retaining it in a latent state, as will be shewn when we consider the subject of heat.

LECTURE XII

Of the Nitrous Acid

Under the head of *liquids* I shall consider *acids* and *alkalis*, though some of them may be exhibited in the form of air, and others in a solid form. These two chemical principles are formed to unite with one another, and then they constitute what is called a *neutral salt*.

Both acids and alkalis are distinguishable by their taste. Another test, and more accurate, is, that acids change the blue juices of vegetables red, and alkalis turn the syrup of violets green.

Acids are generally distinguished according to the three kingdoms to which they belong, viz. *mineral, vegetable,* and *animal.* The mineral acids are three, the *nitrous,* the *vitriolic,* and the *marine.*

The nitrous acid is formed by the union of the purest inflammable air, or the purest nitrous air, with dephlogisticated air. But it is usually procured from nitre by means of the vitriolic acid, which, seizing its base, expels the nitrous acid in a liquid form. On this account this acid is said to be weaker than the vitriolic.

If the nitrous acid be made to pass through a red-hot earthen tube, it will be decomposed, and the greatest part of it be converted into dephlogisticated air.

Like all other acids, the nitrous acid has a strong affinity to water; but it is not capable of so much concentration as the vitriolic. It is generally of an orange or yellow colour; but heat will expel this colour in the form of a red vapour, which is the same acid in the form of air, and loaded with phlogiston; and therefore when it is colourless it is said to be dephlogisticated. But the colourless vapour exposed to heat, or to light, will become coloured again; and the liquid acid imbibing this coloured vapour, becomes coloured as before. This acid tinges the skin of a yellow colour, which does not disappear till the epidermis be changed.

The nitrous acid unites with phlogiston, alkalis, metallic substances, and calcareous earth.

By means of its affinity with phlogiston it occasions that rapid accension called *detonation*, when any salt containing this acid, especially nitre, is applied to hot charcoal, or when charcoal is put to hot nitre. In fact, the charcoal burns so rapidly by means of the dephlogisticated air supplied by the nitre.

A mixture of sulphur assists the accension of these substances, and makes gunpowder, in the explosion of which much nitrous or phlogisticated air is suddenly produced, and expanded by the heat. The application of this force, both to useful and destructive purposes, is well known. If, instead of nitre, a salt made with dephlogisticated marine acid be made use of, the explosion is more easily produced, and is much more violent. Friction will do this as well as heat.

Nitre also enters into the composition of *pulvis fulminans*, viz. three parts nitre, two of dry alkali, and one of sulphur. This composition melts, and yields a blue flame, before it explodes.

By means of the affinity of the nitrous acid to *oil*, another substance containing phlogiston, it is capable of producing not only a great heat, but even a sudden flame, especially when mixed with a little vitriolic acid.

Nitrous acid dissolves all metallic substances except gold and platina, and in the solution nitrous air is produced.

The particular kinds of saline substances formed by the union of the nitrous acid with the several metals and earths may be seen in tables constructed for the purpose. They are all deliquescent.

LECTURE XIII

Of the Vitriolic Acid

The vitriolic acid, so called because it was originally procured from *vitriol*, is now generally procured from sulphur; the dephlogisticated part of the air uniting with it in the act of burning.

That dephlogisticated air is essential to this acid is evident from the decomposition of it; for if the vapour of it be made to pass through a red-hot earthen tube, a great quantity of dephlogisticated air is procured.

This acid has a strong affinity to water, with which it unites with much heat; and it is capable of greater concentration, or of being made specifically heavier, than any other acid. When pure, it is entirely free from colour and smell, owing, probably, to its being free from phlogiston, which is inseparable from the nitrous or marine acids.

The vitriolic acid will dislodge the nitrous, or marine, or any other acid, from their earthy or metallic bases; from which property it is called the strongest of all the acids.

By means of the superior affinity of the vitriolic acid to earths, and especially to *terra ponderosa*, the smallest quantity of it in water may be discovered by a solution of this earth in the marine acid. In this acid the terra ponderosa is held in perfect solution; but with the vitriolic acid it forms a substance that is insoluble in water, and therefore it instantly appears in the form of a white cloud.

Perhaps chiefly from the strong affinity which this acid has with water, *pyrophorus*, consisting of a mixture of alum and several substances containing phlogiston, takes fire spontaneously on exposure to the air. It is commonly made of three parts of alum and one of brown sugar, or of two parts alum, one of salt of tartar, and one of charcoal. They must be heated till they have for some time emitted a vapour that burns with a blue flame.

The saline substances produced by the union of this acid with the several earths and metals, are best exhibited in tables constructed for the purpose. When united to three of the metals, viz. iron, copper, and zinc, they are called *vitriols*, green, blue, and white. And all the substances which this acid unites with crystallize, and do not deliquesce.

This acid unites with oil, and the mixture is always black.

When any substance containing phlogiston is heated in the vitriolic acid, another species of the acid, called *sulphureous*, is formed, of a pungent smell. In reality, it is water impregnated with vitriolic acid air. It makes, however, a distinct species of acid, and is dislodged from its base by most of the others.

Of the Marine Acid

The marine acid is procured from common salt by the vitriolic acid, which unites with its base, the fossil alkali.

This acid is generally of a straw-colour; but this is owing to an impregnation with some earthy matter, most of which it readily dissolves, especially the metallic ones. It is less capable of concentration than the vitriolic or nitrous acids, perhaps from a more intimate union of phlogiston with it. No heat can extract from it any dephlogisticated air.

Though this is denominated a weaker acid than the nitrous, yet it will take silver, lead, or mercury, from their union with the nitrous acid. Upon this principle, a solution of these metals in the nitrous acid will readily discover whether any water contains the marine acid, the latter uniting with the metal dissolved in the former, and forming with it, if it be silver, a *luna cornea*; which being a substance insoluble in water, discovers itself by a cloudy appearance.

The union of the marine acid with earths forms salts that easily deliquesce, but with the metals such as are capable of crystallization; and so also is that formed by the union of this acid to terra ponderosa.

Neither this acid nor the nitrous will dissolve gold or platina; but a mixture of them, called *aqua regia*, will do it.

The marine acid has a strong affinity to dephlogisticated air, and will take it from manganese and other substances; and with this union it becomes a different acid, called *dephlogisticated marine acid*, being water impregnated with dephlogisticated marine acid air, described above.

LECTURE XIV

Of the Vegetable Acids, and others of a less perfect nature

The principal of the vegetable acids are the *acetous* and the *tartareous*. The acetous acid is the produce of a peculiar fermentation of vegetable substances, succeeding the *vinous*, in which ardent spirit it is procured, and succeeded by the *putrefactive*, in which volatile alkali is generated.

Thus wine is converted into vinegar. Crude vinegar, however, contains some ingredient from the vegetable substances from which it was procured: but distillation separates them, and makes the vinegar colourless; though some of the acid is lost in the process.

The acetous acid is concentrated by frost, which does not affect the proper acid, but only the water with which it is united. It may likewise be concentrated by being first combined with alkalies, earths, or metals, and then dislodged by a stronger acid, or by mere heat. Thus the acetous acid, combined with vegetable alkali, forms a substance that is called the *foliated earth of tartar*; and it may be expelled from it by the vitriolic acid. When combined with copper it makes *verdigris*; and from this union heat alone will expel it in a concentrated state. The acetous acid thus concentrated is called *radical vinegar*. Still, however, it is weaker than any of the preceding mineral acids.

Several vegetables, as lemons, sorrel, and unripe fruit, contain acids, ready formed by nature, mixed with some of the essential oil of the plants, which gives them their peculiar flavours. All these acids have peculiar properties; but it is not necessary to note them in this very general view of the subject. Like vinegar, these acids may be concentrated by frost, and also by a combination with other substances, and then expelled by a stronger acid.

The *acid of tartar* is very similar to that of vinegar. Tartar, from which it is procured, is a substance deposited on the inside of wine-casks, though it is also found ready formed in several vegetables. It consists of the vegetable alkali and this peculiar acid. When refined from its impurities, it is called *crystals*, or *cream of tartar*. The acid is procured by mixing the tartar with chalk, or lime, which imbibes the superfluous acid, and this is expelled by the acid of vitriol. Or it may be procured by boiling the tartar with five or six times its weight of water, and then putting the acid of vitriol to it. This unites with the vegetable alkali, and forms vitriolated tartar; and the pure acid of tartar may be procured in crystals, by evaporation and filtration, equal in weight to half the cream of tartar. This acid of tartar is more soluble in water than the cream of tartar.

This acid, united to the mineral alkali, makes *Rochelle salt*.

Every kind of wood, when distilled, or burned, yields a peculiar acid; and it is the vapour of this acid that is so offensive to the eyes in the smoke of wood.

A peculiar acid is obtained from most vegetable substances, especially the farinaceous ones, and from sugar, by distillation with the nitrous acid. This seizes upon the substance with which the acid was united, and especially the phlogiston adhering to it, and then the peculiar *acid of sugar* crystallizes. Thus with three parts of sugar, and thirty of nitrous acid, one part of the proper acid of sugar may be obtained. By the same process an acid may be procured from camphor.

The *bark of oak*, and some other vegetable substances, especially nut-galls, contain a substance which has obtained the name of *the astringent principle*; the peculiar property of which is, that it precipitates solutions of iron in the form of a black powder, and in this manner *ink* is made. But by solution in water and evaporation, crystals, which are a proper *acid of galls*, may be obtained.

Amber is a hard semitransparent substance, chiefly found in Prussia, either dug out of the earth, or thrown up by the sea. It is chiefly remarkable for its electrical property; but by distillation in close vessels there sublimes from it a concreted acid, soluble in 24 times its weight of cold water. Amber seems to be of vegetable origin, and to consist of an oil united to this peculiar acid.

The acids I shall mention next are of a mineral origin; but being of a less perfect nature as acids, I shall only just note them here.

Borax is a substance chiefly found in a crystallized state in some lakes in the East Indies. It consists of the mineral alkali and a peculiar acid, which may be separated, and exhibited in white flakes, by putting acid of vitriol to a solution of it in water. This acid has been called *sedative salt*, from its supposed uses in medicine. It is an acid that requires fifty times its weight of water to dissolve it.

Several other mineral substances, as *arsenic, molybdena, tungsten,* and *wolfram,* in consequence of being treated as the preceding vegetables ones, have been lately found to yield peculiar acids. They are also produced in a concrete state, and require a considerable proportion of water to make them liquid; but as the water in which they are dissolved turns the juice of litmus red, and as they also unite with alkalis, they have all the necessary characteristics of acids.

LECTURE XV

Of the Phosphoric Acid

The most important acid of *animal* origin, though it has lately been found in some mineral substances, is the *phosphoric*.

Phosphorus itself is a remarkable substance, much resembling sulphur, but much more inflammable. It has been procured chiefly, till of late, from urine, but now more generally from *bones*, by means of the vitriolic acid, which unites with the calcareous earth of which bones consist, and sets at liberty the phosphoric acid, or the base of that acid, with which it was naturally combined. The acid thus procured, mixed with charcoal, and exposed to a strong heat, makes phosphorus.

This substance burns with a lambent flame in the common temperature of our atmosphere, but with a strong and vivid flame if it be exposed to the open air when moderately warm. In burning it unites with the dephlogisticated air of the atmosphere, and in this manner the purest phosphoric acid is produced.

This acid is also procured in great purity by means of the nitrous or vitriolic acids, especially the former, which readily combines with the phlogiston of the phosphorus, and thus leaves the acid pure. In this process phlogisticated air is produced.

This acid is perfectly colourless, and when exposed to heat loses all its water, and becomes a glassy substance, not liable to be dissipated by fire, and readily uniting with earths.

United to the mineral alkali, it forms a neutral salt, lately introduced into medicine. United to the mineral and vegetable alkalis naturally contained in urine, it has obtained the name of *microcosmic salt*, frequently used as a flux for mineral substances with a blow-pipe.

Besides the phosphoric, there are other acids of an animal origin; as that of *milk*, that of *sugar of milk*, that of the *animal calculus*, and that of *fat*.

The acid of milk is the sour whey contained in butter-milk, which, by a tedious chemical process, may be obtained pure from any foreign substance.

The sugar of milk is procured by evaporating the whey to dryness, then dissolving it in water, clarifying it with whites of eggs, and evaporating it to the consistence of honey. In this state white crystals of the acid of sugar of milk will be obtained.

By distilling these crystals with nitrous acid, other crystals of the proper *acid of sugar of milk* will be obtained, similar to those of the acid of sugar.

If the human calculus be distilled, it yields a volatile alkali, and something sublimes from it which has a sourish taste, and therefore called the *acid of the calculus*. It is probably some modification of the phosphoric acid.

Animal fat yields an acid by distillation, or by first combining it with quick-lime, and then separating it by the vitriolic acid. Siliceous earth is corroded by this acid.

LECTURE XVI

Of Alkalis

The class of substances that seems particularly formed by nature to unite with acids, and thereby form *neutral salts*, are the *alkalis*. They have all a peculiar acrid taste, not easily defined. They change the blue juices of vegetables green, or purple, and in common with acids have an affinity with water, so as to be capable of being exhibited in a liquid form; though when this water is expelled by heat, some of them will assume a solid form.

Alkalis are of two kinds; the *fixed*, which have no smell, and the *volatile* which have a pungent one.

The fixed alkalis are of *vegetable* or *mineral* origin. When in a solid form, they both melt with a moderate heat, and uniting with earthy substances, make *glass*. With an intense heat they are volatilized.

Vegetable alkali is procured by burning plants, and lixiviating the ashes; a purer kind by the burning of tartar, hence called *salt of tartar*; but the purest of all is got by the deflagration of nitre; the charcoal uniting with the acid as it assumes the form of dephlogisticated air, and the alkali being left behind.

Mineral alkali is found in ashes of sea-weed. It is likewise the basis of sea-salt; from which it is separated by several processes, but especially by the calx of lead, which has a stronger affinity with the marine acid with which it is found combined.

Alkalis united with fixed air are said to be *mild*, and when deprived of it *caustic*, from their readiness to unite with, and thereby *corrode*, vegetable and animal substances. To render them caustic, they are deprived of their fixed air by quick-lime; and in this state they unite with oils, and make *soap*.

Alkalis have a stronger affinity with acids than metals have with them; so that they will precipitate them from their solutions in acid menstruums.

The vegetable fixed alkali has a strong attraction to water, with which it will become saturated in the common state of our atmosphere, when it is said to *deliquesce*; and having the appearance of *oil*, the salt of tartar is thus said to become *oil of tartar per deliquium*. On the other hand, the mineral, or fossil alkali, is apt to lose its water in a dry atmosphere, and then it is said to *effloresce*. In this state it is often found on old walls.

Volatile alkali is procured by burning animal substances; in Egypt (from whence, as contained in *sal ammoniac*, we till of late imported it) from camel's dung; but now from bones, by distillation. To the liquor thus procured they add vitriolic acid, or substances which contain it. This acid unites with the alkali, and common salt being put to it, a double affinity takes place. The vitriolic acid uniting with the mineral alkali of the salt, makes *Glauber salt*, and the marine acid uniting with the volatile alkali, makes *sal ammoniac*. Slaked lime added to this, unites with the marine acid of the ammoniac, and sets loose the volatile alkali in the form of *alkaline air*, which combining with water, makes the liquid caustic volatile alkali. If chalk (containing calcareous earth united with fixed air) be mixed with the sal ammoniac, heat will make the calcareous earth unite with the marine acid, while the fixed air of the chalk will unite with the volatile alkali, and assume a solid form, being the *sal volatile* of the apothecaries.

LECTURE XVII

Of Liquid Inflammable Substances

Of liquid inflammable substances the principal is *spirit of wine*, sometimes called *ardent spirit*, and, when highly rectified, *alcohol*. It is obtained from vegetable substances by their going through the vinous fermentation. It is considerably lighter than water, colourless, and transparent, has a peculiar smell and taste, and the property of inebriating.

Ardent spirit seems to consist of a peculiar combination of phlogiston and water; for when the vapour of it is made to pass through a red-hot earthen tube, it is resolved into water and inflammable air. It is highly inflammable, and burns without smoke, or leaving any residuum; and in the act of burning its phlogiston so unites with dephlogisticated air as to make fixed air.

Ardent spirit mixes readily with water in all proportions, and also with essential oils, and balsams or resins, which are the same thing inspissated.

By its affinity with essential oils, ardent spirit extracts them froth aromatic plants; and these liquors have obtained the name of *tinctures*.

When the tinctures are distilled, the more volatile parts of the essential oils, which come over in distillation, have acquired the name of *waters*; as *Lavender water, Rosemary water*, &c. and what remains in the still is called the *extract* of the plant. If the tinctures be diluted with much water, the resinous part of the plant will be obtained pure, and separated from the extractive part, which will remain dissolved in the water, while the resin separates from it.

Spirit of wine will not dissolve the gummy parts of vegetables; and by this means the gummy substances may be separated from their solutions in water, the spirit uniting with the water only. On the other hand, if resins be dissolved in spirit of wine, the affusion of water will separate them. By means of the affinity of spirit of wine with water, it will seize upon the water in which several salts are dissolved, and thus produce an instant crystallization of them.

Salt of tartar has a greater affinity to water than spirit of wine, and by extracting water from it, will assist in concentrating it; but the best method of rendering spirit wine free from water is distillation, the ardent spirit rising before the water.

Spirit of wine mixed with the vitriolic and other mineral acids, renders them milder, and thereby more proper for certain medicinal uses. This is called *dulcifying* them.

Spirit of wine is a powerful antiseptic, and is therefore of use to preserve vegetable and animal substances from putrefaction.

Of Æther

If spirit of wine be distilled with almost any of the acids, the produce is a liquor which has obtained the name of *Æther*, from its extreme lightness and volatility, being much lighter, and more volatile, than any other fluid that we are acquainted with. It is highly inflammable, but the burning of it is accompanied with smoke, and some soot; and on this account it is a medium between spirit of wine and oil, the acid having taken from the spirit of wine part of the water that was essential to it, at the same time that it communicated something of its acid peculiarly modified; since æthers have different properties according to the acids by which they are made; as the *vitriolic*, the *nitrous*, the *marine*, and the *acetous*. No æther, however, can be made from the marine acid till it has been in some measure dephlogisticated; from which it may be inferred, that dephlogisticated air is necessary to the composition of æther. Vitriolic æther is the most common, in consequence of the process by which it is made being the easiest.

Æther does not mix with water in all proportions, like spirit of wine, but ten parts of water will take up one of æther. It easily mixes with all oils.

It is something remarkable, that though æther will not dissolve gold, it will take from aqua regia the gold that has been previously dissolved in it.

By the quick evaporation of æther a considerable degree of cold may be procured; and on this principle it has sometimes been applied to relieve the head-ach and-other pains.

LECTURE XVIII

Of Oil

Oil is a liquid inflammable substance, of great tenacity, disposed to pour in a stream rather than in drops. It is little, if at all, soluble in water. It burns with smoke and soot, and leaves a residuum of a coaly substance. It consists of acid and water combined with phlogiston.

All oil is the produce of the vegetable or animal kingdom, no proper mineral substance containing any of it.

By distillation oil is in part decomposed, and by this means the thicker kinds of oil are rendered thinner and more volatile, the acid, to which their consistence is chiefly owing, being lost in the process. By repeated distillation it is supposed that all oils may be brought almost to the state of æther, and even of ardent spirit.

Acids act powerfully upon oils, but very differently, according to the nature of each. Alkalies also combine with oils, and the less thin and volatile they are, the more easily are they soluble in alkalies. The union of alkali and oil makes *soap*. All oil dissolves sulphur, and with it makes what is called a *balsam*. Oils also dissolve metallic substances, but most sensibly copper and lead. United with the calx of lead, it is used in painting.

Oil not readily mixing with water, it will diffuse itself over its surface, and, notwithstanding its tenacity, it will do this very rapidly, and to a great extent; and then it has the extraordinary effect of preventing the action of the wind upon the water, so as to prevent the forming of waves. If a quantity of oil and water be put into a glass vessel and swung, the surface of the water below the oil will be seen to change with respect to the vessel, but not that of the oil. If spirit of wine be put upon them, that will be at rest, and both the lower fluids in motion.

Vegetable oil is of two kinds, the *soft*, or *mild*, which has little or no taste or smell, and the *essential* oil, which is thin, and retains the smell and taste of the plant from which it was extracted.

Mild or sweet oil is expressed from the grains or kernels of vegetables, and requires a considerable degree of heat to convert it into vapour, in which state alone it is capable of being inflamed.

Essential oil is volatile in the heat of boiling water, and is generally obtained by means of distillation from the most odoriferous sorts of plants; but is sometimes found in their vesicles, as in the rind of an orange. The strong taste of this kind of oil arises from the disengaged acid which abounds in it; and by this means it is soluble in spirit of wine, which sweet oil is not; but it loses much of this property by repeated distillations. By long exposure to the air it loses its more volatile parts, and thereby approaches to the nature of a resin. This volatile odoriferous principle has been called the *spiritus rector* of the plant.

The essential oils of different plants differ much in their specific gravity, and also in the manner by which they are affected by cold, some being heavier and others lighter than water, and some being more difficultly, and others more easily, congealed. Though the differences with respect to *weight* and *consistency* in these oils is probably owing to the state of the acid that is combined with them, these two properties are wholly independent of each other; some essential oils being very thin and yet heavy, and others thick and yet light. Essential oils are used in perfumes, and also in medicine, acting powerfully the nervous system.

Essential oils are very apt to be adulterated. If it be with sweet oil, it may be discovered by evaporation on white paper, or by a solution in spirit of wine, which will not act upon the sweet oil. If spirit of wine be mixed with it, it will be discovered by a milky appearance upon putting water to it, which uniting with the spirit, will leave the oil much divided. If oil of turpentine, which is the cheapest of essential oils, be mixed with any of the more valuable kinds, it will be discovered by evaporation; a strong smell of turpentine being left on the paper, or cloth, upon which the evaporation was made.

Animal oil, like the vegetable, is of two kinds; the first *butter*, or *fat*, which is easily congealed, owing to the quantity of acid that is intimately combined with it. It resembles the sweet oil of vegetables in having no smell or taste. The other kind of animal oil is extracted by distillation from the flesh, the tendons, the bones, and horns, &c. of animals. It differs essentially from the other kind of animal oil, by containing an alkali instead of an acid. By repeated distillation it becomes highly attenuated and volatile; and in this state it is called the *oil of Dippel*, the discoverer of it.

All oil exposed to much heat is in part decomposed, and acquires a disagreeable smell; and in this state it is said to be *empyreumatic*: but this property is lost by repeated distillations.

Besides the vegetable and animal oils above described, there is a fossil oil called *bitumen*, the several kinds of which differ much in colour and consistence; the most liquid is called *petroleum*, from being found in the cavities of rocks, and the more solid kinds are *amber, jet, asphaltum,* and *pit-coal*. When distilled, the principal component parts of all these substances are an oil and an acid. But all fossil oil is probably of vegetable or animal origin, from masses of vegetables or animals long buried in the earth. Their differences from resins and other oily matters are probably owing to *time;* the combinations of mineral acids and oils so nearly resembling bitumens, the principal difference being their insolubility in spirit of wine.

That the most solid of these, as amber, has been formerly in a liquid state, is evident, from insects and other substances being frequently found in them; and pit-coal has been often found with both the internal texture and external appearance of wood; so that strata of pit-coal have probably been beds of peat in some former state of the earth.

LECTURE XIX

Of Solid Substances

All solid substances are capable of becoming fluid by heat, and most of them may thereby be reduced into a state of vapour, or air; and in passing from a fluid into a solid state their component parts assume a particular mode of arrangement, called *crystallization*, which differs according to the nature of the substance; so that all solids, especially if they be suffered to concrete slowly, may be called *crystals*.

Exclusive of *salts*, which have been considered already, as formed by the union of acids and alkalis, solids in general have obtained the names of *earths*, or *stones*, which differ only in their texture; and they are distinguished into those that are *metallizable*, or those that are not; the former being called *ores*, and the latter simply *earths*; the principal of which are the *calcareous*, *siliceous*, *argillaceous*, *magnesia*, *terra ponderosa*, and a few others which have been discovered lately, but have not been much examined.

Of Calcareous Earth

Calcareous earth is found in the shells of fishes, the bones of animals, chalk, lime-stone, marble, and gypsum: but all calcareous earth is supposed to be of animal origin; and beds of chalk, lime-stone, or marble, are thought to have been beds of shells formed in the sea, in some pristine state of the earth.

The calcareous earth which is found in shells, lime-stone, and marble, is combined with fixed air, discovered by effervescing with acids. To obtain it perfectly pure, the earth must be pounded and washed with water, in order to free it from any saline substance which may be contained in it, then dissolved in distilled vinegar, and precipitated by mild alkalies. Lime-stone exposed to heat loses about half its weight, in fixed air and water, and the remainder, called *quick-lime*, attracts water very powerfully, and their union is attended with much heat, after which it dissolves into a fine powder called *slaked lime*. If it be left exposed to the atmosphere, it will of itself, by gradually imbibing moisture, fall into the state of powder.

Water dissolves about one seven hundreth part of its weight of quicklime, and is then called *lime-water*. Exposed to the air, a crust will be formed on its surface, which is found to consist of calcareous earth and fixed air.

Lime and water mixed with sand make *mortar*, by which means different stones may be made to cohere as one mass, which is the most valuable use of this kind of earth.

Calcareous earth, united with vitriolic acid, makes *gypsum*; and this substance pounded and exposed to heat, parts with its water, and is then called *plaister of Paris*. In this state, by imbibing water again, it becomes a firm substance, and thus is useful in making moulds, &c.

The earth of animal bones is calcareous united to the phosphoric acid.

Of Siliceous Earth

Siliceous earth seems to be formed by nature from chalk, perhaps by the introduction of some unknown acid, which the vitriolic acid is not able to dislodge. It abounds in most substances which are hard enough to strike fire with steel, as *flint, rock crystal*, and most *precious stones*. It is not acted upon by any acid except the fluor and phosphoric, but especially the former: but it is soluble in alkalies; and being then dissolved in water, makes *liquor silicum*, from which the purest siliceous earth may be precipitated by acids. For this purpose about four times the weight of alkali must be made use of. With about equal weights of alkali and siliceous sand is made *glass*, of so great use in admitting light and excluding the weather from our houses, as well as for making various useful utensils. To make glass perfectly colourless, and at the same time more dense, commonly called *flint glass*, manufacturers use a certain proportion of calx of lead and manganese.

Siliceous earth is not affected by the strongest heat, except by means of a burning lens, or dephlogisticated air.

LECTURE XX

Of Argillaceous Earth

Argillaceous earth is found in *clay, schistus,* or *slate,* and in *mica;* but the purest is that which is precipitated from a solution of alum by alkalies; for alum consists of the union of vitriolic acid and argillaceous earth.

This species of earth is ductile with water; it then hardens and contracts by heat, so as to be of the greatest use in forming *bricks,* or stones of any required form or size. By means of the property of clay to contract in the fire, Mr. Wedgwood has constructed an excellent thermometer to measure the degrees of extreme heat.

The ductility of clay seems to depend upon some acid, probably the vitriolic, adhering to it; for it loses that property when it is burned into a brick, but recovers it when it has been again dissolved in an acid.

Of Terra Ponderosa

Terra ponderosa, or *marmor metallicum,* is generally found in two states, viz. united to vitriolic acid, when it is called *calk,* or to fixed air, when it is called *terra ponderosa aerata.*

To obtain it pure from its union with the vitriolic acid, it must be melted with about twice its weight of fixed alkali; which unites with the acid, and forming a saline substance, may be washed out of it. In this state it contains water, and therefore, when exposed to heat, will yield fixed air; whereas the terra ponderosa aerata will not yield fixed air by heat only, but when steam is made to pass over it when red hot. This proves that water is essential to the composition of fixed air.

This stone is distinguishable by its great specific gravity, being four times as heavy as water; but though in this it resembles an *ore,* it has not been found to be metallizable.

Of Magnesia

This species of earth is found in *steatites,* or *soap rock, Spanish chalk, asbestus,* and *Muscovy talck;* but the purest is got by dissolving *Epsom salts* (which consists of this earth united to the vitriolic acid) and precipitating it by a mild alkali. In this state it becomes united to fixed air, which may be expelled by heat. It is then *calcined*, or *caustic*, but differs from quick-lime by not being soluble in water.

Asbestus, which contains much of this kind of earth, is remarkable for not being destructible by heat, though it is sometimes found in flexible fibres, so as to be capable of being woven into cloth.

Muscovy talck is remarkable for the thin and transparent flakes into which it is divisible, and thereby capable of various uses.

There are some other distinct species of earth, particularly one brought from Botany Bay, and another called *Stontiate*, from the place where it was found in Scotland; but they have not as yet been much examined.

All stones formed by nature are compounded, and to distinguish them from one another, and ascertain the parts of which they consist, is the subject of *lithology*, a very extensive branch of knowledge.

All the simple earths are nearly, if not absolutely, *infusible*; but when they are mixed they may all be fused.

LECTURE XXI

Of Ores

Metallizable earths, commonly called *ores*, when united to phlogiston, make the metals, distinguishable for their specific gravity, their opacity, shining appearance, and fusibility.

All the proper metals are *malleable*, and those which are not so are called *semi-metals*.

The metals again are subdivided into the *perfect* and *imperfect*. The former, which are *gold, silver*, and *platina*, suffer no change by fusion, or the longest continued heat: whereas heat calcines or dissipates the phlogiston of the imperfect metals, which are *mercury, lead, copper, iron*, and *tin*, so that they return to the state of earth; and this earth is always heavier than the metal, though of less specific gravity, having received an addition of weight from water or air: but these earths, or ores, being exposed to heat in contact with substances containing phlogiston, again become metals, and are then said to be *revived*.

The semi-metals are *bismuth, zinc, nickel*, regulus of *arsenic*, of *cobalt*, of *antimony*, of *manganese*, of *wolfram*, and of *molybdena*.

All metallic substances are crystallizable, and each in a peculiar form, which is discovered by leaving a hole in the bottom of the crucible in which they are melted, and drawing out the stopper, when the mass is beginning to lose its fluidity.

Some of the metals will not unite to others when hot, and others of them will; and such as will unite with others are called *solders*. Thus tin is a solder for lead, and brass, gold, or silver, for iron.

Ores are never found in regular strata, like the different kinds of earth; but in places which have formerly been cavities, running in all directions, with respect to the regular strata, and commonly called *veins*.

Many of the ores in their natural state are said to be *mineralized* with arsenic or sulphur, those substances being intimately united with the metallic earths.

In order to convert the ores into metals, some of them are first reduced to powder, to wash out the earthy or saline particles. They are then kept in a red heat, which the workmen call *roasting*, in order to drive away the arsenic, or sulphur, which are volatile; and in the last place they are fused in contact with charcoal, or other substances containing phlogiston; and to promote the fusion, lime-stone is frequently mixed with them. When the operation is completed, the unmetallic parts are converted into glass, or *scoria*, which lies on the surface, whereas the metal is found at the bottom.

To discover the quantity of metal in a small piece of ore is called *assaying*.

When metals are fused together, the specific gravity, fusibility, and other properties are changed, and in such a manner as could not be discovered from the properties of the constituent parts.

Of Gold

Gold is the heaviest of all metallic bodies except platina. It appears yellow or reddish by reflected light, but green or blue by transmitted light, when it is reduced to thin plates.

Though gold undergoes no change in a common furnace, or burning lens, it may, in part, at least, be calcined by the electric shock.

Gold has the greatest *ductility*, and in wires of equal diameters, it has the greatest *tenacity*, of all the metals. One grain of it may be made to cover 56 square inches; some gold leaf being less than a 200,000th part of an inch thick; and when it is made to cover a silver wire, the gold upon it may not be more than one twelfth part of the thickness of the gold leaf.

This metal is soluble in aqua regia; and being precipitated by a volatile alkali, makes a powder called *aurum fulminans*, which is one fourth heavier than the gold, and explodes with great violence in a heat something greater than that of boiling water.

Tin precipitates gold in the form of a purple powder, called the *powder of Cassius*, from the inventor of it, and is used in enamels, or the glassy coating which is given to metals by heat.

Gold unites with most of the metals, especially with mercury, and these mixtures are called *amalgams*. In gilding, the amalgam is applied to the surface of the metal to be gilded, and the mercury is driven off by heat, leaving the gold attached to the surface.

Gold mixed with iron, makes it harder, for the purpose of cutting instruments.

To separate gold from the imperfect metals, such as copper, &c. it is mixed with lead, and then exposed to a strong heat, which calcines the lead, and with it the imperfect metals, leaving the gold pure. This process is called *cupellation*, from being performed in a small crucible called a *cupell*. When the gold is mixed with silver, three parts more of silver are put to it, and then the silver is dissolved by nitrous acid, leaving the gold pure. This process is called *quartation*, from the gold being one fourth part of the mass.

The fineness of gold is generally estimated by dividing the gold into twenty-four parts, called *carats*. The phrase twenty-three carats fine means that the mass contains twenty-three parts out of twenty-four of pure gold, the remainder being *alloy*, of some baser metal. The fineness of gold may in some measure be discovered by the colour it leaves upon a *touch-stone*, or fine-grained basaltes.

Gold is generally found nearly pure, but mixed with earth, or diffused in fine grains through stones.

LECTURE XXII

Of Silver

Silver is the whitest of all the metals, very ductile, but less so than gold; the thinnest leaves of it being one third thicker than those of gold. It is not calcined in the heat of a common furnace, but partially so by repeated fusion, or a strong burning lens.

Sulphureous fumes unite with silver, and tinge it black. The nitrous acid dissolves it, and will hold more than half its weight of it in solution. When fully saturated, this solution deposits crystals, which are called *lunar nitre*, or *nitre of silver*. When these crystals are melted, and the water they contain driven off, a black substance, called *lapis infernalis*, or *lunar caustic*, is formed. This is used as a cautery in surgery. A strong heat will decompose this lunar nitre, and recover the silver.

Though the nitrous acid dissolves silver the most readily, the marine acid will deprive the nitrous of it, and form a substance called *luna cornea*, because, when it is melted and cold, it becomes a transparent mass something resembling *horn*. From this luna cornea the purest silver may be obtained. The vitriolic acid will likewise deprive the nitrous of the silver contained in it, and form a white powder, not easily soluble in water.

A fulminating silver may be made by the following process: the silver must first be dissolved in pale nitrous acid, then precipitated by lime-water, dried, and exposed to the air three days. It must then be washed in caustic volatile alkali, after which the fluid must be decanted, and the black powder left to dry in the air. The slightest friction will cause this powder to fulminate. It is said, that even a drop of water falling upon it will produce this effect; so that it ought to be made only in very small quantities, and managed with the greatest caution.

Most of the metals precipitate silver. That by mercury may be made to assume the form of a tree, called *arbor Dianæ*.

Silver is found native in Peru; and the ores frequently contain sulphur, or arsenic, or both.

Of Platina

Platina is a metal lately discovered in the gold mines of Mexico, where it is found in small particles, never exceeding the size of a pea, mixed with ferruginous sand and quartz.

The strongest fire will not melt these grains, though it will make them cohere; but they may be melted by a burning lens, or a blow-pipe supplied with dephlogisticated air.

Pure platina is the heaviest body in nature, its specific gravity exceeding twenty-two. It is very malleable, though considerably harder than gold or silver, and has the property of welding in common with iron. This metal is not affected by exposure to the air, or by any simple acid, though concentrated and hot; but it is dissolved by dephlogisticated marine acid, and by aqua regia, in which a little nitrous air is procured. The solution is brown, and when diluted yellow. This liquor is very corrosive, and tinges animal substances of a blackish brown colour. Platina is precipitated from a solution in aqua regia by sal-ammoniac, as gold is by martial vitriol. Iron is precipitated from this solution by the Prussian alkali. Also most of the metals precipitate platina, but not in its metallic state.

Arsenic facilitates the solution of platina; and by melting it with equal parts of arsenic and vegetable alkali, and then reducing the mass to a powder, it may be made to take any form; and a strong heat will dissipate the arsenic and the alkali, leaving the platina in the shape required, not fusible by any heat in a common furnace.

Platina does not readily combine with gold or silver, and it resists the action of mercury as much as iron; but it mixes well with lead, making it less ductile, and even brittle, according to the proportion of the platina. With copper it forms a compound which takes a beautiful polish, not liable to tarnish, and is therefore used with advantage for mirrors of reflecting telescopes. It unites easily with tin, and also with bismuth, antimony, and zinc.

LECTURE XXIII

Of Mercury

Mercury is the most fusible of all the metals, not becoming solid but in 40° below 0 in Fahrenheit's thermometer. It is then malleable. It is heavier than any other metal except gold or platina. It is volatile in a temperature much lower than that of boiling water, and in vacuo in the common temperature of the atmosphere; and at six hundred it boils.

In a degree of heat in which it would rise easily in vapour, mercury imbibes pure air, and becomes a red calx, called *precipitate per se*. At a greater degree of heat it parts with that air, and is running mercury again.

Mercury is not perceptibly altered by exposure to the air.

Mercury is acted upon by the vitriolic acid when hot. In this process vitriolic acid air is procured, and the mercury is converted into a white substance, which being dipped in water becomes yellow, called *turbith mineral*, one third heavier than the mercury from which it was made. By heat this substance parts with its pure air, and becomes running mercury; but if the process be made in a clean earthen vessel, there will remain a portion of *red calx*, which cannot be reduced by any degree of heat, except in contact with some substance containing phlogiston. If this be done with a burning lens, in inflammable air, much of the air will be absorbed.

Mercury is dissolved most readily in the nitrous acid, when the purest nitrous air is procured; and there remains a substance which is first yellow, and by continuance red, called *red precipitate*. In a greater degree of heat the dephlogisticated air will be recovered, and the mercury be revived; but the substance yields nitrous air after it becomes solid, and till it changes from yellow to red.

The precipitates of mercury from acids by means of alkalies possess the property of exploding, when they are exposed to a gradual heat in an iron spoon, after having been triturated with one sixth of their weight of the flowers of sulphur. The residuum consists of a violet-coloured powder, which, by sublimation, is converted into cinnabar.

It seems, therefore, as if the sulphur combined suddenly with the mercury, and expelled the dephlogisticated air in an elastic state.

The marine acid seizes upon mercury dissolved in nitrous acid, and if the acid be dephlogisticated, the precipitate is *corrosive sublimate*; but with common marine acid, it is called *calomel*, or *mercurius dulcis*. This preparation is generally made in the dry way, by triturating equal parts of mercury, common salt and vitriol, and exposing the whole to a moderate heat; when the corrosive sublimate rises, and adheres to the upper part of the glass vessel in which the process is made.

Mercury combines readily with sulphur by trituration, and with it forms a black powder called *Ethiops mineral*. A more intimate combination of mercury and sulphur is made by means of fire. This is called *cinnabar*, about one third of which is sulphur. Vermillion is cinnabar reduced to powder.

Mercury readily unites with oil, and with it forms a deep black or blue compound, used in medicine.

It readily combines with most of the metals, and when it is used in a sufficient quantity to make them soft, the mixture is called an *amalgam*. It combines most readily with gold, silver, lead, tin, bismuth, and zinc. Looking-glasses are covered on the back with an amalgam of mercury and tin.

When mercury is united with lead or other metals, it is rendered less brilliant and less fluid; but agitation in pure air converts the impure metal into a calx, together with much of the mercury, in the form of a black powder.

Heat recovers the pure air, and the mercury, leaving the calx of the impure metal.

Much mercury is found native in a slaty kind of earth, or in masses of clay or stone; but the greatest quantity is found combined with sulphur in *native cinnabar*.

LECTURE XXIV

Of Lead

Lead is a metal of a bluish tinge, of no great tenacity, but very considerable specific gravity, being heavier than silver. It melts long before it is red hot, and is then calcined, if it be in contact with respirable air. When boiling it emits fumes, and calcines very rapidly. It may be granulated by being poured into a wooden box, and agitated. During congelation it is brittle, so that the parts will separate by the stroke of a hammer; and by this means the form of its crystals may be discovered.

In the progress of calcination it first becomes a dusky grey powder, then yellow, when it is called *massicot*; then, by imbibing pure air, it becomes red, and is called *minium*, or *red lead*. In a greater degree of heat it becomes massicot again, having parted with its pure air. If the heat be too great, and applied rapidly, it becomes a flaky substance, called *litharge*; and with more heat it becomes a *glass*, which readily unites with metallic calces and earths, and is a principal ingredient in the manufacture of *flint glass*, giving it its peculiar density and refractive power.

Though lead soon tarnishes, the imperfect calx thus made does not separate from the rest of the metal, and therefore protects it from any farther action of the air, by which means it is very useful for the covering of houses, and other similar purposes. All acids act upon lead, and form with it different saline substances. *White-lead* consists of its union with vinegar and pure air. Also dissolved in vinegar, and crystallized, it becomes *sugar of lead*, which, like all the other preparations of this metal, is a deadly poison.

Oils dissolve the calces of lead, which, by this means, is the basis of paints, plaisters, &c.

By means of heat litharge decomposes common salt, the lead uniting with the marine acid, and forming a yellow substance, used in painting, and by this means the fossil alkali is separated.

Lead unites with most metals, though not with iron. Two parts of lead and one of tin make a *solder*, which melts with less heat than either of the metals separately; but a composition of eight parts of bismuth, five of lead, and three of tin, makes a metal which melts in boiling water.

This metal will be dissolved by water if it contain any saline matter, and the drinking of it occasions a peculiar kind of cholic.

Lead is sometimes found native, but generally minerally mineralized with sulphur or arsenic, and often mixed with a small quantity of silver.

Of Copper

Copper is a metal of a reddish or brownish colour, considerably sonorous, and very malleable.

At a heat far below ignition, the surface, of copper becomes covered with a range of prismatic colours, the commencement of its calcination; and with more heat a black scale is formed, which easily separates from the metal, and in a strong heat it melts, and burns with a bluish green flame.

Copper rusts by exposure to the air; but the partially-calcined surface adheres to the metal, as in the case of lead, and thus preserves it from farther corrosion.

Copper dissolved in the vitriolic acid forms crystals of a blot colour, called *blue copperas*. From this solution it is precipitated by iron, which by this means becomes coated with copper. The nitrous acid dissolves copper with most rapidity, producing nitrous air. If the solution be distilled, almost all the acid will be retained in the residuum, which is white; but more heat will expel the acid, chiefly in the form of dephlogisticated air, and the remainder will be a black substance, consisting of the pure calx of copper. The vegetable acids dissolve copper as well as the mineral ones, which makes the use of this metal for culinary purposes in some cases dangerous. To prevent this they give it a coat of tin. The solution of copper in the vegetable acid is called *verdigris*.

Alkalies dissolve copper as well as acids. With the volatile alkali a blue liquor is formed, but in some cases it becomes colourless. All the circumstances of this change of colour have not yet been examined. Both oil and sulphur will dissolve copper, and with the latter it forms a blackish grey compound, used by dyers.

Copper readily unites with melted tin, at a temperature much lower than that which is necessary to melt the copper; by which means copper vessels are easily covered with a coating of tin. A mixture of copper and tin, called *bronze,* the specific gravity of which is greater than that of the medium of the two metals, is used in casting statues, cannon, and bells; and in a certain proportion this mixture is excellent for the purpose of mirrors of reflecting telescopes, receiving a fine polish, and not being apt to tarnish.

Copper and arsenic make a brittle compound called *tombach*; and with zinc it makes the useful compound commonly called *brass*, in which zinc is about one third of its weight.

Copper is sometimes found native; but commonly mixed with sulphur, in ores of a red, green, or blue colour.

Copper being an earlier discovery than that of iron, was formerly used for weapons and the shoeing of horses; and the ancients had a method, with which we are not well acquainted, of giving it a considerable degree of hardness, so that a sword made of it might have a pretty good edge.

LECTURE XXV

Of Iron

Iron is a metal of a bluish colour, of the greatest hardness, the most variable in its properties, and the most useful of all the metals; so that without it it is hardly possible for any people to make great advances in arts and civilization.

This metal readily parts with its phlogiston, so as to be very subject to calcine, or rust, by exposure to the air. The same is evident by the colours which appear on its surface when exposed to heat, and also when it is struck with flint; the particles that fly from it being iron partially calcined. In consequence of its readily parting with its phlogiston, it is capable of burning, like wood or other fuel, in pure air.

Iron and platina have the property of *welding* when very hot, so that two pieces may be joined without any solder.

When iron is heated in contact with steam, part of the water takes the place of the phlogiston, while the rest unites with it, and makes inflammable air. By this means the iron acquires one third more weight, and becomes what is called *finery cinder*. This substance, heated in inflammable air, imbibes it, parts with its water, and becomes perfect iron again. If the iron be heated in pure air, it also imbibes the water, of which that air chiefly consists, and also a portion of the peculiar element of the pure air.

The solution of iron in spirit of vitriol produces *green copperas*; which being calcined, becomes a red substance, called *colcothar*.

The precipitate of iron, by an infusion of galls, is the colouring matter in *ink*, which is kept suspended by means of gum. The precipitate from the same solution by phlogisticated alkali, is *Prussian blue*.

Water saturated with fixed air dissolves iron, and makes a pleasant chalybeat.

The calx of iron gives a green colour to glass.

Iron readily combines with sulphur. When they are found combined by nature, the substance is called *pyrites*.

The union of phosphoric acid with iron makes it brittle when cold, commonly called *cold short*; and the union of arsenic makes it brittle when hot, thence called *red short*.

Iron unites with gold, silver, and platina, and plunged in a white heat into mercury, it becomes coated with it; and if the process be frequently repeated, it will become brittle, which shews that there is some mutual action between them.

Iron readily unites with tin; and by dipping thin plates of iron into melted tin, they get a complete coating of it, and make the *tinned plates* in common use.

When crude iron comes from the smelting furnace it is brittle; and when it is white within, it is extremely hard; but when it has a black grain, owing to its having more phlogiston, it is softer, and may be filed and bored.

Cast iron becomes *malleable* by being exposed to a blast of air when nearly melting; the consequence of which is a discharge of inflammable air, and the separation of a liquid substance, which, when concreted, is called *finery cinder*. The iron generally loses one fourth of its weight in the process. Crude iron contains much *plumbago*, and the access of pure air probably assists in discharging it, by converting it into air, chiefly inflammable.

Malleable iron, exposed to a red heat in contact with charcoal, called *cementation*, converts it into *steel*, which has the properties of becoming much harder than iron, and very elastic, by being first made very hot, and then suddenly cooled, by plunging it in cold water. By first making it very hard, and then giving different degrees of heat, and cooling it in those different degrees, it is capable of a great variety of *tempers*, proper for different uses. Of the degrees of heat workmen judge by the change of colour on its surface. Steel, like crude iron, is capable of being melted without losing its properties. It is then called *cast steel*, and is of a more uniform texture. Iron acquires some little weight by being converted into steel; and when dissolved in acid, it yields more plumbago. Steel has something less specific gravity than iron. If the cementation be continued too long, the steel acquires a darkish fracture, it is more fusible, and incapable of welding. Steel heated in contact with earthy matters, is reduced to iron.

Iron is the only substance capable of *magnetism*; and hardened steel alone is capable of retaining magnetism. The loadstone is an ore of iron.

LECTURE XXVI

Of Tin

Tin is a metal of a slightly yellowish cast, though harder than lead, very malleable, but of no great tenacity; so that wires cannot be made of it. It easily extends under the hammer, and plates of it, called *tinfoil*, are made only one thousandth part of an inch thick, and might be made as thin again.

Tin has less specific gravity than any other metal. It melts long before ignition, at 410 of Fahrenheit, and by the continuance of heat is slowly converted into a white powder, which is the chief ingredient in *putty*, used in polishing, &c. Like lead, it is brittle when heated little short of fusion, and may be reduced into grains by agitation as it passes from a fluid to a solid state.

The calx of tin resists fusion more than that of any other metal, which makes it useful in making an opaque white enamel.

Tin loses its bright surface when exposed to the air, but is not properly subject to rust; so that it is useful in protecting iron and other metals from the effects of the atmosphere.

Concentrated vitriolic acid, assisted by heat, dissolves half its weight of tin, and yields vitriolic acid air. With more water it yields inflammable air. During the solution the phlogiston of the tin uniting with the acid, forms sulphur, which makes it turbid. By long standing, or the addition of water, the calx of tin is precipitated from the solution. The nitrous acid dissolves tin very rapidly without heat, and yields but little nitrous air. With marine acid this metal yields inflammable air. With aqua regia it assumes the form of a gelatinous substance used by dyers to heighten the colour of some red tinctures, so as to produce a bright scarlet in dying wool.

A transparent liquor, which emits very copious fumes, called, from the inventor, *the smoking liquor of Libavius*, is made by distilling equal parts of amalgam of tin and mercury with corrosive sublimate, triturated together. A colourless liquor comes over first, and then a thick white fume, which condenses into the transparent liquor above mentioned. Mr. Adet has shewn, that this liquor bears the same relation to the common solution of tin, that corrosive sublimate does to calomel, and has given an ingenious solution of many of its properties.

Tin detonates with nitre; and if the crystals made by the solution of copper in the nitrous acid be inclosed in tinfoil, nitrous fumes will be emitted, and the whole will become red hot. Also if five times its weight of sulphur be added to melted tin, a black brittle compound, which readily takes fire, will be formed.

Another combination of tin, sulphur, and mercury, makes a light yellow substance called *aurum musivum* used in painting.

Tin is the principal ingredient in the composition of *pewter*, the other ingredients being lead, zinc, bismuth, and copper; each pewterer having his peculiar receipt. It is also used in coating copper and iron plates, and in silvering looking-glasses, besides being cast into a variety of forms, when it is called *block tin*.

Tin is sometimes found native, but is generally mineralized with sulphur and arsenic. The latter is thought to be always contained in tin, and to be the cause of the crackling noise made by bending plates of it.

Of the Semi-metals

Bismuth is a semi-metal of a yellowish or reddish cast, but little subject to change in the air; harder than lead, but easily broken, and reducible to powder. When broken it exhibits large shining facets, in a variety of positions. Thin pieces of it are considerably sonorous.

Bismuth melts at about 460° of Fahrenheit. With more heat it ignites, and burns with a slight blue flame, while a yellowish calx, called *flowers of bismuth*, is produced. With more heat it becomes a greenish glass. In a strong heat, and in close vessels, this metal sublimes.

Vitriolic acid, even concentrated and boiling, has but little effect upon bismuth; but the nitrous acid acts upon it with the greatest rapidity and violence, producing much nitrous air, mixed with phlogisticated nitrous vapour. From the solution of bismuth in this acid, a white substance, called *magistery of bismuth*, is precipitated by the affusion of water. This has been used as a paint for the skin but has been thought to injure it.

The marine acid does not readily act upon bismuth; but when concentrated, it forms with it a saline combination, which does not easily crystallize, but may be sublimed in the form of a soft fusible salt, called *butter of bismuth*.

Bismuth unites with most metallic substances, and in general renders them more fusible. When calcined with the imperfect metals, it unites with them, and has the same effect as lead in cupellation.

Bismuth is used in the composition of pewter, in printers' types, and other metallic mixtures.

This metal is sometimes found native, but more commonly mineralized with sulphur.

LECTURE XXVII

Of Nickel

Nickel is a semi-metal of a reddish cast, of great hardness, and always magnetical; on which account it is supposed to contain iron, though chemists have not yet been able to separate them.

The purest nickel was so infusible as not to run into a mass in the strongest heat of a smith's forge; but then it was in some degree malleable.

Concentrated acid of vitriol only corrodes nickel. Alkalies precipitate it from its solution in the nitrous acid, and dissolve the precipitate. It readily unites with sulphur.

Nickel is found either native or mineralized with several other metals, especially with copper, when it is called *kupfer nickel*, or *false copper*, being of a reddish or copper colour.

This semi-metal has not yet been applied to any use.

Of Arsenic

What is commonly called *arsenic* is the calx of a semi-metal called the *regulus of arsenic*. It is a white and brittle substance, expelled from the ores of several metals by heat. It is then refined by a second sublimation, and melted into the masses in which it is commonly sold. This calx is soluble in about eighty times its weight of cold water, or in fifteen times its weight of boiling water. It acts in many respects like an acid, as it decomposes nitre by distillation, when the nitrous acid flies off, and the *arsenical salt of Macquer* remains behind.

When the calx of arsenic is distilled with sulphur, the vitriolic acid flies off, and a substance of a yellow colour, called *orpiment*, is produced. This appears to consist of sulphur and the regulus of arsenic; part of the sulphur receiving pure air from the calx, to which it communicates phlogiston; and thus the sulphur becomes converted into vitriolic acid, while the arsenical calx is reduced, and combines with the rest of the sulphur.

The combination of sulphur and arsenic, by melting them together, is of a red colour, known by the name of *realgal*, or *realgar*. It is less volatile than orpiment.

The solution of fixed alkali dissolves the calx of arsenic, and by means of heat a brown tenacious mass is produced, and having also a disagreeable smell, it is called *liver of arsenic*.

The regulus of arsenic is of a yellow colour, subject to tarnish or grow black, by exposure to the air, very brittle, and of a laminated texture. In close vessels it wholly sublimes, but burns with a small flame in pure air.

Vitriolic acid has little action upon this semi-metal, except when hot; but the nitrous acid acts readily upon it, and likewise dissolves the calx, as does boiling marine acid, though it affects it very little when cold.

Most of the metals unite with the regulus of arsenic.

Dephlogisticated marine acid converts the calx of arsenic into *arsenical acid* by giving it pure air.

The acid of arsenic acts more or less upon all metals, but the phenomena do not appear to be of much importance.

The calx of acid is used in a variety of the arts, especially in the manufactory of glass. Orpiment and realgar are used as pigments. Some attempts have been made to introduce it into medicine, but being dangerous, the experiments should be made with caution.

Of Cobalt

Cobalt is a semi-metal of a grey or steel colour, of a close-grained fracture, more difficult of fusion than copper, not easily calcined. It soon tarnishes in the air, but water has no effect upon it.

Cobalt, dissolved in *aqua regia*, makes an excellent sympathetic ink, appearing green when held to the fire, and disappearing when cold, unless it has been heated too much, when it burns the paper.

The calx of cobalt is of a deep blue colour, which, when fused, makes the blue glass called *smalt*. The ore of cobalt, called *zaffre*, is found in several parts of Europe, but chiefly in Saxony. As it is commonly sold, it contains twice or thrice its weight of powder of flints. The smalt is usually composed of one part of calcined cobalt, fused with two parts of powder of flint and one of pot-ash.

The chief use of cobalt is for making smalt; but the powder and the blue-stone used by laundresses is a preparation made by the Dutch of a coarse kind of smalt.

Of Zinc

Zinc is a semi-metal of a bluish cast, brighter than lead, and so far malleable as not to be broken by a hammer, though it cannot be much extended. When broken by bending, it appears to consist of cubical grains. If it be heated nearly to melting, it will be sufficiently brittle to be pulverized. It melts long before ignition, and when it is red hot, it burns with a dazzling white flame, and is calcined with such rapidity, that its calx flies up in the form of white flowers, called *flowers of zinc*, or *philosophical wool*. In a stronger heat they become a clear yellow glass. Heated in close vessels, this metal rises without decomposition, being the most volatile of all the metals except the regulus of *arsenic*.

Zinc dissolved in diluted vitriolic acid, yields much inflammable air, and has a residuum, which appears to be plumbago, and the liquor forms crystals, called *white copperas*. This metals also yields inflammable air when dissolved in the marine acid. Dissolved in the nitrous acid, it yields dephlogisticated nitrous air, with very little proper nitrous air.

The ore of zinc, called *calamine*, is generally of a white colour; and the chief use of it is to unite it with copper, with which it makes brass and other gold-coloured mixtures of metals. The calx and the salts of this metal are occasionally used in medicine.

LECTURE XXVIII

Of Antimony

The regulus of antimony is of a silvery white colour, of a scaly texture, very brittle, and melts soon after ignition. By continuance of heat it calcines in white fumes, called *argentine flowers of antimony*, which melt into a hyacinthine glass. In close vessels it rises without decomposition. Its calx is soluble in water, like that of arsenic. This metal tarnishes, but does not properly rust, by exposure to the air.

This metal is soluble in aqua regia. It detonates with nitre, and what remains of equal parts of nitre and regulus of antimony after detonation, in a hot crucible, is called *diaphoretic antimony*. The water used in this preparation contains a portion of the calx suspended by the alkali, and being precipitated by an acid, is called *ceruse of antimony*.

When regulus of antimony is pulverized and mixed with twice its weight of corrosive sublimate (which is attended with heat) and then distilled with a gentle fire, a thick fluid comes over, which is congealed in the receiver, or in the neck of the retort, and is called *butter of antimony*. The residuum consists of revived mercury, with some regulus and calx of antimony. When this butter of antimony is thrown into pure water, there is a white precipitate, called *powder of algaroth*, a violent emetic. Nitrous acid dissolves the butter of antimony; and when an equal weight of nitrous acid has been three times distilled to dryness from butter of antimony, the residuum, after ignition, is called *bezoar mineral*, and seems to be little more than a calx of the metal.

Crude antimony, which has been much used in the experiments of alchemists, is a combination of sulphur and regulus of antimony. Heat melts it, and finally converts it into glass, of a dark red colour, called *liver of antimony*. If antimony be melted or boiled with a fixed alkali, a precipitate is made by cooling, called *kermes mineral*, formerly used in medicine. The antimonial preparations that are now most in use are *antimonial wine* and *tartar emetic*. The wine is made by infusing pulverized glass of antimony in Spanish wine some days, and filtering the clear fluid through paper. The emetic tartar, or antimonial tartar, is a saline substance, composed of acid of tartar, vegetable alkali, and antimony partially calcined. The preparation may be seen in the Dispensaries.

The regulus of antimony is used in the form of pills, which purge more or less in proportion to the acid they meet with; and as they undergo little or no change in passing through the body, they are called *perpetual pills*.

Of Manganese

Manganese is a hard, black mineral, very ponderous, and the regulus of it is a semi-metal of a dull white colour when broken, but soon grows dark by exposure to the air. It is hard and brittle, though not pulverizable, rough in its fracture, and of very difficult fusion. Its calces are white when imperfect, but black, or dark green, when perfect. The white calx is soluble in acids. When broken in pieces, it falls into powder by a spontaneous calcination, and this powder is magnetical, though the mass was not possessed of that property. The black calx of manganese is altogether insoluble in acids. It contains much dephlogisticated air.

The calx of manganese is used in making glass; the glass destroying the colour of that of the other materials, and thereby making the whole mass transparent.

This semi-metal mixes with most of the metals in fusion, but not with mercury.

There is another ore of manganese, called *black woad*, which inflames spontaneously when mixed with oil.

Of Wolfram

Wolfram is a mineral of a brownish or black colour, found in the tin mines of Cornwall, of a radiated or foliated texture, shining almost like a metal. It contains much of the calx of manganese, and iron; but when the substance is pulverized, these are easily dissolved, and the calx of wolfram is found to be yellow.

This calx turns blue by exposure to light; and an hundred grains of it heated with charcoal will yield sixty grains of a peculiar metal, in small particles, which, when broken, look like steel. It is soluble in the vitriolic or marine acids, and reduced to a yellow calx by nitrous acid or aqua regia.

Of Molybdena

Molybdena is a substance which much resembles plumbago; but its texture is scaly, and not easily pulverized, on account of a degree of flexibility which its laminæ possess. With extreme heat, and mixed with charcoal, it yields small particles of a metal that is grey, brittle, and extremely infusible; and uniting with several of the metals, it forms with them brittle or friable compounds. By heat it is converted into a white calx.

Of Solid Combustible Substances

There yet remains a class of solid substances, of the *combustible* kind, but most of them have been already considered under the form of the fluids, from which they are originally formed, as *bitumen*, *pit-coal*, and *amber*; or under the principal ingredients of which they are composed, as *sulphur* and *plumbago*.

There only remains to be mentioned the *diamond*, which is of a nature quite different from that of the other precious stones, the principal ingredient in which is siliceous earth, which renders them not liable to be much affected by heat. On the contrary, the diamond is a combustible substance; for in a degree of heat somewhat greater than that which will melt silver, it burns with a slight flame, diminishes common air, and leaves a soot behind. Also, if diamond powder be triturated with vitriolic acid, it turns it black, which is another proof of its containing phlogiston.

The diamond is valued on account of its extreme hardness, the exquisite polish it is capable of, and its extraordinary refractive power; for light falling on its interior surface with an angle of incidence greater than 24½ will be wholly reflected, whereas in glass it requires an angle of 41 degrees.

LECTURE XXIX

Of the Doctrine of Phlogiston and the Composition of Water

It was supposed to be a great discovery of Mr. Stahl, that all inflammable substances, as well as metals, contain a principle, or substance, to which he gave the name of phlogiston, and that the addition or deprivation of this substance makes some of the most remarkable changes in bodies, especially that the union of a metallic calx and this substance makes a metal; and that combustion consists in the separation of phlogiston from the substances that contain it. That it is the same principle, or substance, that enters into all inflammable substances, and metals, is evident, from its being disengaged from any of them, and entering into the composition of any of the others. Thus the phlogiston of charcoal or inflammable air becomes the phlogiston of any of the metals, when the calx is heated in contact with either of them.

On the contrary, Mr. Lavoisier and most of the French chemists, are of opinion, that there is no such principle, or substance, as phlogiston; that metals and other inflammable bodies are simple substances, which have an affinity to pure air; and that combustion consists not in the separation of any thing from the inflammable substance, but in the union of pure air with it.

They moreover say, that water is not, as has been commonly supposed, a simple substance, but that it consists of two elements, viz. pure air, or *oxygene*, and another, to which they give the name of *hydrogene*, which, with the principle of *heat*, called by them *calorique*, is inflammable air.

The principal fact adduced by them to prove that metals do not lose any thing when they become calces, but only gain something, is, that mercury becomes a calx, called *precipitate per se*, by imbibing pure air, and that it becomes running mercury again by parting with it.

This is acknowledged: but it is almost the only case of any calx being revived without the help of some known phlogistic substance; and in this particular case it is not absurd to suppose, that the mercury, in becoming precipitate per se, may retain all its phlogiston, as well as imbibe pure air, and therefore be revived by simply parting with that air. In many other cases the same metal, in different states, contains more or less phlogiston, as cast iron, malleable iron, and steel. Also there is a calx of mercury made by the acid of vitriol, which cannot be revived without the help of inflammable air, or some other substance supposed to contain phlogiston: and that the inflammable air is really imbibed in these processes, is evident, from its wholly disappearing, and nothing being left in the vessel in which the process is made beside the metal that is revived by it. If precipitate per se be revived in inflammable air, the air will be imbibed, so that running mercury may contain more or less phlogiston.

The antiphlogistians also say, that the diminution of atmospherical air by the burning of phosphorus is a proof of their theory; the pure air being imbibed by that substance, and nothing emitted from it. But there is the same proof of phosphorus containing phlogiston, that there is of dry flesh containing it; since the produce of the solution of it in nitrous acid, and its effect upon the acid, are the same, viz. the production of phlogisticated air, and the phlogistication of the acid.

Their proof that water is decomposed, is, that in sending steam over hot iron, inflammable air (which they suppose to be one constituent part of it) is procured; while the other part, viz. the oxygene, unites with the iron, and adds to its weight. But it is replied, that the inflammable air may be well supposed to be the phlogiston of the iron, united to part of the water, as its base, while the remainder of the water is imbibed by the calx; and that it is mere water, and not pure air, or oxygene, that is retained in the iron, is evident, from nothing but pure water being recovered when this calx of iron is revived in inflammable air, in which case the inflammable air wholly disappears, taking the place of the water, by which it had been expelled.

In answer to this it is said, that the pure air expelled from the calx uniting with the inflammable air in the vessel, recomposes the water found after this process. But in every other case in which any substance containing pure air is heated in inflammable air, though the inflammable air be in part imbibed, some *fixed air* is produced, and this fixed air is composed of the pure air in the substance and part of the inflammable air in the vessel. Thus, if *minium*, which contains pure air, and *massicot*, which contains none, be heated in inflammable air, in both the cases lead will be revived by the absorption of inflammable air; but in the former case only, and not in the latter, will fixed air be produced. The calx of iron, therefore, having the same effect with massicot, when treated in the same manner, appears to contain no more pure air than massicot does.

Besides this explanation of the facts on which the new theory is founded, which shews it to be unnecessary, the old hypothesis being sufficient for the purpose, some facts are alledged, as inconsistent with the new doctrine.

If the calx of iron made by water, and charcoal made by the greatest degree of heat, be mixed together, a great quantity of inflammable air will be produced; though, according to the new theory, neither of these substances contained any water, which they maintain to be the only origin of it. But this fact is easily explained upon the doctrine of phlogiston; the water in this calx uniting with the phlogiston of the charcoal, and then forming inflammable air; and it is the same kind of inflammable air that is made from charcoal and water.

Also the union of inflammable and pure air, when they are fired together by means of the electric spark, produces not pure water, as, according to the new theory, it ought to do, but *nitrous acid*.

To this it has been objected, that the acid thus produced came from the decomposition of phlogisticated air, a small portion of which was at first contained in the mixture of the two kinds of air. But when every particle of phlogisticated air is excluded, the strongest acid is procured.

They find, indeed, that by the slow burning of inflammable air in pure air, they get pure water. But then it appears, that whenever this is the case, there is a production of phlogisticated air, which contains the necessary element of nitrous acid; and this is always the case when there is a little surplus of the inflammable air that is fired along with the pure air, as the acid is always procured when there is a redundancy of pure air.

That much water should be procured by the decomposition of these kinds of air, is easily accounted for, by supposing that water, or steam, is the basis of these, as well as of all other kinds of air.

Since air something better than that of the atmosphere is constantly produced from water by converting it into vapour, and also by removing the pressure of the atmosphere, and these processes do not appear to have any limits; it seems probable, that *water* united to the principle of *heat*; constitutes atmospherical air; and if so, it must consist of the elements of both dephlogisticated and phlogisticated air; which is a supposition very different from that of the French chemists.

LECTURE XXX

Of Heat

Heat is an affection of bodies well known by the sensation that it excites. It is produced by friction or compression, as by the striking of flint against steel, and the hammering of iron, by the reflection or refraction of light, and by the combustion of inflammable substances.

It has been long disputed, whether the cause of heat be properly a *substance*, or some particular affection of the particles that compose the substance that is heated. But be it a substance, or a principle of any other kind, it is capable of being transferred from one body to another, and the communication of it is attended with the following circumstances. All substances are expanded by heat, but some in a greater degree than others; as metals more than earthy substances, and charcoal more than wood. Also some receive and transmit heat through their substance more readily than others; metals more so than earths, and of the metals, copper more readily than iron. Instruments contrived to ascertain the expansion of substances by heat, are called *pyrometers*, and are of various constructions.

As a standard to measure the degrees of heat, mercury is in general preferable to any other substance, on account of its readily receiving, and communicating, heat through its whole mass. *Thermometers*, therefore, or instruments to measure the degrees of heat, are generally constructed of it, though, as it is subject to become solid in a great degree of cold, ardent spirit, which will not freeze at all, is more proper in that particular case.

The graduation of thermometers is arbitrary. In that of Fahrenheit, which is chiefly used in England, the freezing point of water is 32°, and the boiling point 212°. In that of Reaumur, which is chiefly used abroad, the freezing point of water is 0, and the boiling point 80. To measure the degrees of heat above ignition, Mr. Wedgwood has happily contrived to use pieces of clay, which contract in the fire; and he has also been able to find the coincidence of the degrees in mercurial thermometers with those of his own.

To measure the degrees of heat and cold during a person's absence, Lord George Cavendish contrived an instrument, in which a small bason received the mercury, that was raised higher than the place for which it was regulated by heat or cold, without a power of returning. But Mr. Six has lately hit upon a better method, viz. introducing into the tube of his thermometer a small piece of iron, which is raised by the ascent of the mercury, and prevented from descending by a small spring; but which may be brought back to its former place by a magnet acting through the glass.

Heat, like light, is propagated in right lines; and what is more remarkable, cold observes the same laws. For if the substance emitting heat without light, as iron below ignition, be placed in the focus of a burning mirror, a thermometer in the focus of a similar mirror, placed parallel to it, though at a considerable distance, will be heated by it, and if a piece of ice be placed there, the mercury will fall.

Heat assists the solvent power of almost all menstrua; so that many substances will unite in a certain degree of heat, which will form no union at all without it, as dephlogisticated and inflammable air.

If substances be of the same kind, they will receive heat from one another, in proportion to their masses. Thus, if a quantity of water heated to 40° be mixed with another equal quantity of water heated to 20°, the whole mass will be heated to 30°. But if the substances be of different kinds, they will receive heat from each other in different proportions, according to their *capacity* (as it is called) of receiving heat. Thus, if a pint of mercury of the temperature of 136 be mixed with a pint of water of the temperature of 50, the temperature of the two after mixture will not be a medium between those two numbers, viz. 93, but 76; consequently the mercury was cooled 60°, while the water was heated only 26; so that 26 degrees of heat in water correspond to 60 in mercury. But mercury is about 13 times specifically heavier than water, so that an equal weight of mercury would contain only one thirtieth part of this heat; and dividing 26 by 13, the quotient is 2. If *weight*, therefore, be considered, the heat discovered by water should be reckoned as 2 instead of 60; and consequently when water receives 2 degrees of heat, an equal weight of mercury will receive 60°; and dividing both the numbers by 2, if the heat of water be 1, that of the mercury will be 30. Or since they receive equal degrees of heat, whether they discover it or not (and the less they discover, the more they retain in a latent state) a pound of mercury contains no more than one thirtieth part of the heat actually existing in a pound of water of the same temperature. Water, therefore, is said to have a greater capacity for receiving and retaining heat, without discovering it, than mercury, in the proportion of 30 to 1, if weight be considered, or of 60 to 26, that is of 30 to 13; if *bulk* be the standard, though, according to some, it is as 3 to 2.

The capacity of receiving heat in the substance is greatest in a state of vapour, and least in that of a solid; so that when ice is converted into water, heat is absorbed, and more still when it is converted into vapour; and on the contrary, when vapour is converted into water, it gives out the heat which it had imbibed, and when it becomes ice it gives out still more.

If equal quantities of ice and water be exposed to heat at the temperature of 32°, the ice will only become water, without receiving any additional sensible heat; but an equal quantity of water in the same situation would be raised to 178°, so that 146 degrees of heat will be imbibed, and remain in latent in the water, in consequence of its passing from a state of ice: and heat communicated by a given weight of vapour will raise an equal weight of a nonevaporable substance, of the same capacity with water, 943 degrees; so that much more heat is latent in steam, than in the water from which it was formed.

This doctrine of latent heat explains a great variety of phænomena in nature; as that of cooling bodies by evaporation, the vapour of water, or any other fluid substance, absorbing and carrying off the heat they had before.

Water, perfectly at rest, will fall considerably below the freezing point, and yet continue fluid: but on the slightest agitation, the congelation of the whole, or part of it, takes place instantly, and if the whole be not solid, it will instantly rise to 32°, the freezing point. From whatever cause, some motion seems necessary to the commencement of congelation, at least in a moderate temperature; but whenever any part of the water becomes solid, it gives out some of the heat it had before, and that heat which was before latent becoming sensible, and being diffused through the whole mass, raises its temperature.

On the same principle, when water heated higher than the boiling point in a digester is suddenly permitted to escape in the form of steam, the remainder is instantly reduced to the common boiling point, the heat above that point being carried off in a latent state by the steam.

Had it not been for this wise provision in nature, the whole of any quantity of water would, in all cases of freezing, have become solid at once; and also the whole of any quantity that was heated to the point of boiling, would have been converted into steam at once; circumstances which would have been extremely inconvenient, and often fatal.

This doctrine also explains the effect of freezing mixtures, as that of salt and snow. These solid substances, on being mixed, become fluid, and that fluid absorbing much heat, deprives all the neighbouring bodies of part of what they had. But if the temperature at which the mixture is made be as low as that to which this mixture would have brought it, it has no effect, and in a lower temperature this new fluid would become solid; for that mixture has only a certain determinate capacity for heat, and if the neighbouring bodies have less heat, they will take from it.

It has been observed, that the comparative heat of bodies containing phlogiston is increased by calcination or combustion; so that the calx of iron has a greater capacity for heat, and therefore contains more latent heat, than the metal.

In general it is not found, that the same substances have their capacity for receiving heat increased by an increase of temperature; but this is said to be the case with a mixture of ardent spirit and water, and also that of spirit of vitriol and water.

Since all substances contain a greater or less quantity of heat, and in consequence of being deprived of it become colder and colder, it is a question of some curiosity to determine the extent to which this can go, or at what degree in the scale of a thermometer any substance would be absolutely cold, or deprived of all heat; and an attempt has been made to solve this problem in the following manner. Comparing the capacity of water with that of ice, by means of a third substance, viz. mercury, it has been found, that if that of ice be 9°, that of water is 10°; so that water in becoming ice gives out one tenth part of its whole quantity of heat. But it has been shown, that ice in becoming water absorbs 146 degrees of heat. This, therefore, being one tenth part of the whole heat of water, it must have contained 1460 degrees; so that taking 32 degrees, which is the freezing point, from that number, the point of absolute cold will be 1426 below 0 of Fahrenheit's scale.

By a computation, made by means of the heat of inflammable and dephlogisticated air, at the temperature of 50, Dr. Crawford finds, that it contains nearly 1550 degrees of heat; so that the point of absolute cold will be 1500 below 0. But more experiments are wanted to solve this curious problem to entire satisfaction.

LECTURE XXXI

Of Animal Heat

Since all animals, and especially those that have red blood, are much hotter than the medium in which they live, the source of this heat has become the subject of much investigation; and as the most probable theory is that of Dr. Crawford, I shall give a short detail of the reasons on which it is founded.

Having, with the most scrupulous attention, ascertained the *latent*, or, as he calls it, the *absolute* heat of blood, and also that of the aliments of which it is composed, he finds that it contains more than could have been derived from *them*. Also finding that the absolute heat of arterial blood exceeds that of venous blood, in the proportion of 11½ to 10, he concludes that it derives its heat from the air respired in the lungs, and that it parts with this *latent* heat, so that it becomes sensible, in the course of its circulation, in which it becomes loaded with phlogiston, which it communicates to the air in the lungs.

That this heat is furnished by the *air*, he proves, by finding, that that which we inspire contains more heat than that which we expire, or than the aqueous humor which we expire along with it, in a very considerable proportion; so that if the heat contained in the pure air did not become latent in the blood, it would raise its temperature higher than that of red-hot iron. And again, if the venous blood, in being converted into arterial blood, did not receive a supply of latent heat from the air, its temperature would fall from 96 to 104 below 0 in Fahrenheit's thermometer.

That the heat procured by combustion has the same source, viz. the dephlogisticated air that is decomposed in the process, is generally allowed; and Dr. Crawford finds, that when equal portions of air are altered by the respiration of a Guinea pig, or by the burning of charcoal, the quantity of heat communicated by the two processes is nearly equal.

The following facts are also alleged in favour of his theory. Whereas animals which have much red blood, and respire much, have the power of keeping themselves in a temperature considerably higher than that of the surrounding atmosphere, other animals, as *frogs* and *serpents*, are nearly of the same temperature with it; and those animals which have the largest respiratory organs, as birds, are the warmest; also the degree of heat is in some measure proportionable to the quantity of air that is respired in a given time, as in violent exercise.

It has been observed, that animals in a medium hotter than the blood have a power of preserving themselves in the same temperature. In this case the heat is probably carried off by perspiration, while the blood ceases to receive, or give out, any heat; and Dr. Crawford finds, that when an animal is placed in a warm medium the colour of the venous blood approaches nearer to that of the arterial than when it is placed in a colder medium; and also, that it phlogisticates the air less than in the former case; so that in these circumstances respiration has not the same effect that it has in a colder temperature, in giving the body an additional quantity of heat; which is an excellent provision in nature, as the heat is not wanted, but, on the contrary, would prove inconvenient.

LECTURE XXXII

Of Light

Another most important agent in nature, and one that has a near connexion with heat, is *light*, being emitted by all bodies in a state of ignition, and especially by the sun, the great source of light and of heat to this habitable world.

Whether light consists of particles of matter (which is most probable) or be the undulation of a peculiar fluid, filling all space, it is emitted from all luminous bodies in right lines.

Falling upon other bodies, part of the light is *reflected* at an angle equal to that of its incidence, though not by impinging on the reflecting surface, but by a power acting at a small distance from it. But another part of the light enters the body, and is *refracted* or bent *towards*, or *from*, the perpendicular to the surface of the new medium, if the incidence be oblique to it. In general, rays of light falling obliquely on any medium are bent as if they were attracted by it, when it has a greater density, or contains more of the inflammable principle, than the medium through which it was transmitted to it. More of the rays are reflected when they fall upon a body with a small degree of obliquity to its surface, and more of them are transmitted, or enter the body, when their incidence is nearer to a perpendicular.

The velocity with which light is emitted or reflected is the same, and so great that it passes from the sun to the earth in about eight minutes and twelve seconds.

Rays of light emitted or reflected from a body entering the pupil of the eye, are so refracted by the humours of it, as to be united at the surface of the retina, and so make images of the objects, by means of which they are visible to us; and the magnifying power of telescopes or microscopes depends upon contriving, by means of reflections or refractions, that pencils of rays issuing from every point of any object shall first diverge, and then converge, as they would have done from a much larger object, or from one placed much nearer to the eye.

When a beam of light is bent out of its course by refraction, all the rays of which it consists are not equally refracted, but some of them more and others less; and the colour which they are disposed to exhibit is connected invariably with the degree of their refrangibility; the red-coloured rays being the least, and the violet the most refrangible, and the rest being more or less so in proportion to their nearness to these, which are the extremes, in the following order, violet, indigo, blue, green, yellow, orange, red.

These colours, when separated as much possible, are still contiguous; and all the shades of each colour have likewise their separate and invariable degrees of refrangibility. When separated as distinctly as possible, they divide the whole space between them exactly as a musical chord is divided in order to found the several notes and half notes of an octave.

These differently-coloured rays of light are also separated in passing through the transparent medium of air and water, in consequence of which the sky appears blue and the sea green, these rays being returned, while the red ones proceed to a greater distance. By this means also objects at the bottom of the sea appear to divers red, and so do all objects enlightened by an evening sun.

The mixture of all the differently-coloured rays, in the proportions in which they cover the coloured image above mentioned, makes a *white*, and the absence of all light is *blackness*.

By means of the different refrangibility of light, the colours of the rainbow may be explained.

The distance to which the differently-coloured rays are separated from each other is not in proportion to the mean refractive power of the medium, but depends upon the peculiar constitution of the substance by which they are refracted. The *dispersing power* of glass, into the composition of which *lead* enters, is great in proportion to the mean refraction; and it is proportionally little in that glass in which there is much alkaline salt. The construction of *achromatic telescopes* depends upon this principle.

Not only have different rays of light these different properties with respect to bodies, so as to be more or less refracted, or dispersed, by them, but different sides of the same rays seem to have different properties, for they are differently affected on entering a piece of *island crystal*. With the same degree of incidence; part of the pencil of rays, consisting of all the colours, proceeds in one direction, and the rest in a different one; so that objects seen through a piece of this substance appear double.

At the surface of all bodies rays of light are promiscuously reflected, or transmitted.

But if the next surface be very near to it, the rays of one colour chiefly are reflected, and the rest transmitted, and these places occur alternately for rays of each of the colours in passing from the thinnest to the thickest parts of the medium; so that several series, or orders, of colours will be visible on the surface of the same thin transparent body. On this principle coloured rings appear between a plane and a convex lens, in a little oil on the surface of water, and in bubbles made with soap and water.

When rays of light pass near to any body, so as to come within the sphere of its attraction and repulsion, an *inflection* takes place; all the kinds of rays being bent *towards*, or *from*, the body, and these powers affecting some rays more than others, they are by this means also separated from each other, so that coloured streaks appear both within the shadow, and the outside of it, the red rays being inflected at the greatest distance from the body.

Part of the light which enters bodies is retained within them, and proceeds no farther; but so loosely in some kinds of bodies, that a small degree of heat is sufficient to expel it again, so as to make the body visible in the dark: but the more heat is applied, the sooner is all the light expelled. This is a strong argument for the materiality of light. *Bolognian phosphorus* is a substance which has this property; but a composition made by Mr. Canton, of calcined oyster-shells and sulphur, in a much greater degree. However, white paper, and most substances, except the metals, are possessed of this property in a small degree.

Some bodies, especially phosphorus, and animal substances tending to putrefaction, emit light without being sensibly hot.

The *colours* of vegetables, and likewise their *taste* and *smell*, depend upon light. It is also by means of light falling on the leaves and other green parts of plants, that they emit dephlogisticated air, which preserves the atmosphere fit for respiration.

It is light that imparts colour to the skins of men, by means of the fluid immediately under them. This is the cause of *tanning*, of the *copper colour* of the North Americans, and the *black* of the Negroes. Light also gives colour to several other substances, especially the solutions of mercury in acids.

LECTURE XXXIII

Of Magnetism

Magnetism is a property peculiar to iron, or some ores of it. The earth itself, owing probably to the iron ores contained in it, has the same property. But though all iron is acted upon by magnetism, *steel* only is capable of having the power communicated to it.

Every magnet has two poles, denominated *north* and *south*, each of which attracts the other, and repels that of the same kind with itself. If a magnet be cut into two parts, between the two poles, it will make two magnets, the parts that were contiguous becoming opposite poles.

Though the poles of a magnet are denominated *north* and *south*, they do not constantly, and in all parts of the earth, point due north or south, but in most places to the east or west of them, and with a considerable variation in a course of time. Also a magnet exactly balanced at its center will have a declination from an horizontal position of about 70 degrees. The former is called the *variation*, and the latter the *dipping* of the magnetic needle.

A straight bar of iron which has been long fixed in a vertical position, will become a magnet, the lower end becoming a north pole, and the upper end a south one; for if it be suspended horizontally, the lower end will point towards the north, and the upper end towards the south. Also any bar of iron, not magnetical, held in a vertical position, will become a temporary magnet, the lower end becoming a north pole, and the upper end a south one; and a few strokes of a hammer will fix the poles for a short time, though the position of the ends be changed. Magnetism may likewise be given to a bar of iron by placing it firmly in the position of the dipping-needle, and rubbing it hard one way with a polished steel instrument. Iron will also become magnetical by ignition and quenching it in water in the position of the dipping-needle.

Magnetism acts, without any diminution of its force, through any medium; and iron not magnetical will have that power while it is in connexion with a magnet, or rather the power of the magnet is extended through the iron.

Steel filings gently thrown upon a magnet, adhere to it in a curious manner; and the filings, acquiring magnetism by the contact, adhere together, and form a number of small magnets, which arrange themselves according to the attraction of the poles of the original magnet. This experiment is made to the most advantage upon a piece of pasteboard, or paper, placed over the magnet.

Magnetism is communicated by the friction, or the near position, of a magnet to a piece of steel of a size less than it. For this reason a combination of magnetical bars will have a greater effect than a single one; and in the following manner, beginning without any magnetism at all, the greatest quantity may be procured. Six bars of steel may be rendered slightly magnetical by fixing each of them successively to an upright poker, and stroking it several times from the bottom to the top with the lower end of an old pair of tongs. If then four of these bars be joined, the magnetism of the remaining two will be much increased by a proper method of rubbing with them; and by changing their places, joining the strongest, and acting upon the weakest, they may all be made as magnetical as they are capable of being.

The strength of a natural magnet may be increased by covering its polar extremities with steel. This is called the *arming* of the loadstone.

To account for the variation of the needle, Dr. Halley supposed the earth to consist of two parts, an external *shell* and an internal *nucleus*, detached, and having a revolution distinct from it; and that the action of the poles of the shell and of the nucleus would explain all the varieties in the position of the needle. But others think that the cause of the magnetism of the earth is not *within*, but *without* itself. One reason for this opinion is, that a magnet is liable to be affected by a strong aurora borealis; and another is, that the variation of the needle proceeds in such manner as supposes that the motion of the nucleus must be quicker than that of the shell of the earth; whereas, since it is most natural to suppose that motion was communicated to the nucleus by the shell, it would be slower.

Some idea of the quantity and the progress of the variation of the needle may be formed from the following facts.—At the Cape of Good Hope, when it was discovered by the Portuguese, in 1486, there was no variation, the needle there pointing due north; in 1622 it was about 2 degrees westward, in 1675 it was 8° W. in 1700 about 11° W. in 1756 about 18° W. and in 1774 about 21½° W. In London, in 1580, the variation was 11 degrees 15 seconds E.; in 1622 it was 6° E. in 1634 it was 4 deg. 5 min. E. in 1657 it was nothing at all; in 1672 it was 2 deg. 30 min. W. in 1692 it was 6 deg. W. in 1753 it was about 16 W. and at present it is about 21 W.

The longitude may in some places be found by the variation of the needle; and Mr. Churchman, of America, having given much attention to the subject, comparing the observations of others, and many of his own, thinks that he has found a method of determining the longitude to a great degree of certainty, in most cases, by this means.

He says there are two magnetic poles of the earth, one to the north and the other to the south, at different distances from the poles of the earth, and revolving in different times; and from the combined influence of these two poles he deduces rules for the position of the needle in all places of the earth, and at all times, past, present, or to come.

The north magnetic pole, he says, makes a complete revolution in 426 years, 77 days, 9 hours, and the south pole in about 5459 years. In the beginning of the year 1777 the north magnetic pole was in 76 deg. 4 min. north latitude; and in longitude from Greenwich 140 deg. east; and the south was in 72 deg. south latitude, and 140 deg. east from Greenwich.

LECTURE XXXIV

Of Electricity

Electricity is a property belonging to, or capable of being communicated to, all substances whatever; and whereas by some of them it is transmitted with great ease, and by others with much difficulty, they have been divided into two classes, and denominated *conductors* or *non-conductors* of electricity. Also the latter receiving this power by friction, and other means, are termed *electrics*, and the former *non-electrics*.

Metals of all kinds, and water, are conductors, though in very different degrees; so also is charcoal. All other substances, and also a perfect vacuum, are non-conductors of electricity. But many of these substances, when they are made very hot, as glass, resin, baked wood, and perhaps all the rest on which the experiment can be made in this state, are conductors.

It is the property of all kinds of electrics, when they are rubbed by bodies different from themselves, to attract light substances of all kinds, to exhibit an appearance of *light*, attended with a particular *sound*, on the approach of any conductor; and if the nostrils are presented, they are affected with a *smell* like that of phosphorus. This attraction is most easily explained by supposing that electricity is produced by a fluid exceedingly elastic, or repulsive of itself, and attracted by all other substances.

An electric exhibiting the appearances above mentioned, is said to be *excited*, and some of them, particularly the *tourmaline*, are excited by heating and cooling, as well as by friction. It appears, however, that excitation consists in the mere transferring of electricity from one substance to another, and that the great source of electricity is in the earth. On this account it is necessary to the considerable excitation of any electric, that the substance against which it is rubbed (hence termed *the rubber*) have a communication with the earth, by means of conductors; for if the rubber be *insulated*, that is cut off from all communication with the earth by means of electrics, the friction has but little effect.

When insulated bodies have been attracted by, and brought into contact with, an excited electric, they begin to be repelled by it, and also to repel one another; nor will they be attracted again till they have been brought into contact with some conductor communicating with the earth; but after this they will be attracted as at first.

If conductors be *insulated*, electric powers may be communicated to them by the approach of excited electrics, or the contact of other electrified bodies. They will then attract light bodies, and give sparks, &c. like the excited electrics themselves.

When electricity is strongly communicated to insulated animal bodies, the pulse is quickened, and perspiration increased; and if they receive, or part with, their electricity on a sudden, a painful sensation is felt at the place of communication. But what is more extraordinary, is, that the influence of the brain and nerves upon the muscles seems to be of an electric nature.

This is one of the last and most important of all philosophical discoveries. I shall, therefore, give the result of all the observations that have hitherto been made on the subject, in a *series of propositions*, drawn up by an intelligent friend, who has given much more attention to it than I have done.

1. The nerve of the limb of an animal being laid bare, and surrounded with a piece of sheet lead, or of tinfoil, if a communication be formed between the nerve thus armed and any of the neighbouring muscles, by means of a piece of zinc, strong contractions will be produced in the limb.

2. If a portion of the nerve which has been laid bare be armed as above, contractions will be produced as powerfully, by forming the communication between the armed and bare part of the nerve, as between the armed part and muscle.

3. A similar effect is produced by arming a nerve and simply touching the armed part of the nerve with the metallic conductor.

4. Contractions will take place if a muscle be armed, and a communication be formed by means of the conductor between it and a neighbouring nerve. The same effect will be produced if the communication be formed between the armed muscle and another muscle, which is contiguous to it.

5. Contractions may be produced in the limb of an animal by bringing the pieces of metal into contact with each other at some distance from the limb, provided the latter make part of a line of communication between the two metallic conductors.

The experiment which proves this is made in the following manner. The amputated limb of an animal being placed upon a table, let the operator hold with one hand the principal nerve, previously laid bare, and in the other let him hold a piece of zinc; let a small plate of lead or silver be then laid upon the table, at some distance from the limb, and a communication be formed, by means of water, between the limb and the part of the table where the metal is lying. If the operator touch the piece of silver with the zinc, contractions will be produced in the limb the moment that the metals come into contact with each other. The same effect will be produced if the two pieces of metal be previously placed in contact, and the operator touch one of them with his finger. This fact was discovered by Mr. William Cruikshank.

6. Contractions can be produced in the amputated leg of a frog, by putting it into water, and bringing the two metals into contact with each other at a small distance from the limb.

7. The influence which has passed through, and excited contractions in, one limb, may be made to pass through, and excite contractions in, another limb. In performing this experiment it is necessary to attend to the following circumstances: let two amputated limbs of a frog be taken; let one of them be laid upon a table, and its foot be folded in a piece of silver; let a person lift up the nerve of this limb with a silver probe, and another person hold in his hand a piece of zinc, with which he is to touch the silver including the foot; let the person holding the zinc in one hand catch with the other the nerve of the second limb, and he who touches the nerve of the first limb is to hold in his other hand the foot of the second; let the zinc now be applied to the silver including the foot of the first limb, and contractions will immediately be excited in both limbs.

8. The heart is the only involuntary muscle in which contractions can be excited by these experiments.

9. Contractions are produced more strongly, the farther the coating is placed from the origin of the nerve.

10. Animals which were almost dead have been found to be considerably revived by exciting this influence.

11. When these experiments are repeated upon an animal that has been killed by opium, or by the electric shock, very slight contractions are produced; and no contractions whatever will take place in an animal that has been killed by corrosive sublimate, or that has been starved to death.

12. Zinc appears to be the best exciter when applied to gold, silver, molybdena, steel, or copper. The latter metals, however, excite but feeble contractions when applied to each other. Next to zinc, in contact with these metals, tin and lead, and silver and lead, appear to be the most powerful exciters.

At least two kinds of fishes, the *torpedo* and the *electrical eel,* have a voluntary power of giving so strong a shock to the water in which they swim, as to affect fishes and other animals which come near them; and by a conducing communication between different parts of these fishes, an electric shock may be given exactly like that of the Leyden phial, which will be described hereafter; and if the communication be interrupted, a flash of electric light will be perceived.

The growth of vegetables is also quickened by electricity.

LECTURE XXXV

The same Subject continued

No electric can be excited without producing electric appearances in the body with which it is excited, provided that body be insulated; for this insulated rubber will attract light bodies, give sparks, and make a snapping noise, upon the approach of a conductor, as well as the excited electric.

If an insulated conductor be pointed, or if a pointed conductor, communicating with the earth, be held pretty near it, little or no electric appearance will be exhibited, only a light will appear at each of the points during the act of excitation, and a current of air will be sensible from off them both.

The effect of pointed bodies is best explained on the supposition of the electric matter in one body repelling that in another; and consequently the electricity belonging to a body with a large surface making a greater resistance to the entrance of foreign electricity than that belonging to a smaller.

These two electricities, viz. that of the excited electric, and that of the rubber, though similar to, are the reverse of, one another. A body attracted by the one will be repelled by the other, and they will attract, and in all respects act upon, one another more sensibly than upon other bodies; so that two pieces of glass or silk possessed of contrary electricities will cohere firmly together, and require a considerable force to separate them.

These two electricities having been first discovered by producing one of them from glass, and the other from amber, sealing-wax, sulphur, rosin, &c. first obtained the names of *vitreous* and *resinous* electricity; and it being afterwards imagined that one of them was a redundancy, and the other a deficiency, of a supposed electric fluid, the former has obtained the name of *positive*, and the latter that of *negative*, electricity; and these terms are now principally in use.

Positive and negative electricity may be distinguished from each other by the manner in which they appear at the points of bodies. From a pointed body electrified positively, there issues a stream of light, divided into denser streams, at the extremities; whereas, when the point is electrified negatively, the light is more minutely divided, and diffused equally. The former of these is called a *brush*, and the latter a *star*.

If a conductor not insulated be brought within the atmosphere (that is the sphere of action) of any electrified body, it acquires the electricity opposite to that of the electrified body, and the nearer it is brought, the stronger opposite electricity does it acquire, till the one receive a spark from the other, and then the electricity of both will be discharged.

The electric substance which separates the two conductors possessing these two opposite kinds of electricity, is said to be *charged*. Plates of glass are the most convenient for this purpose, and the thinner the plate the greater is the charge it is capable of holding. The conductors contiguous to each side of the glass are called their *coating*.

Agreeably to the above-mentioned general principle, it is necessary that one side of the charged glass have a communication with the rubber, while the other receives the electricity from the conductor, or with the conductor, while the other receives from the rubber.

It follows also, that the two sides of the plate thus charged are always possessed of the two opposite electricities; that side which communicates with the excited electric having the electricity of the electric, and that which communicates with the rubber, that of the rubber.

There is, consequently, a very eager attraction between these two electricities with which the different sides of the plate are charged, and when a proper communication is made by means of conductors, a flash of electric light, attended with a report (which is greater or less in proportion to the quantity of electricity communicated to them, and the goodness of the conductors) is perceived between them, and the electricity of both sides is thereby discharged.

The substance of the glass itself in, or upon, which these electricities exist, is impervious to electricity, and does not permit them to unite; but if they be very strong, and the plate of glass very thin, they will force a passage through the glass. This, however, always breaks the glass, and renders it incapable of another charge.

The flash of light, together with the explosion between the two opposite sides of a charged electric, is generally called the *electric shock*, on account of the disagreeable sensation it gives any animal whose body is made use of to form the communication been them.

The electric shock is always found to perform the circuit from one side of the charged glass to the other by the shortest passage through the best conductors. Common communicated electricity also observes the same rule in its transmission from one body to another.

It has not been found, that the electric shock takes up any sensible space of time in being transmitted to the greatest distances.

The electric shock, as also the common electric spark, displaces the air through which it passes; and if its passage from conductor to conductor be interrupted by non-conductors of a moderate thickness, it will rend and tear them in its passage, in such a manner as to exhibit the appearance of a sudden expansion of the air about the center of the shock.

If the electric circuit be interrupted, the electric matter, during the discharge, will pass to any other body that lies near its path, and instantly return. This may be called the *lateral explosion*. The effect of this lateral explosion through a brass chain, when the quantity of electricity is very great, will be the discolouring and partial burning of the paper on which it lies.

If a great quantity of electricity be accumulated, as in a *battery*, the explosion will pass over the surfaces of imperfect conductors without entering them, and the effect will be a strong *concussion* of the substance. Also the electric matter thus accumulated and condensed will, by its repulsion, form *concentric circles*, which will appear by melting the surface of a flat piece of metal on which the explosion is received.

If an electric shock, or strong spark, be made to pass through, or over, the belly of a muscle, it forces it to contract, as in a convulsion.

If a strong shock be sent through a small animal body, it will often deprive it instantly of life.

When the electric shock is very strong, it will give polarity to magnetic needles, and sometimes it reverses their poles.

Great shocks, by which animals are killed, are said to hasten putrefaction.

Electricity and lightning are in all respects the same thing; since every effect of lightning may be imitated by electricity, and every experiment in electricity may be made with lightning, brought down from the clouds by means of insulated pointed rods of metal.

LECTURE XXXVI

The same Subject continued

Three curious and important instruments, which are among the latest improvements in electricity, deserve a particular explanation, and in all of them the effect depends upon the general principles mentioned above, viz. that bodies placed within the influence, or, as it is usually termed, within the atmosphere, of an electrified body, are affected by a contrary electricity, and that these two electricities mutually attract each other. These instruments are the *electrophorus*, the *condenser* of electricity, and the *doubler* of it.

The electrophorus consists of an insulated conducting plate applied to an insulated electric. If the latter have any electricity communicated to it, for example the negative, the positive electricity of the former will be attracted by it, and consequently the plate will be capable of receiving electricity from any body communicating with the earth; being, in this situation, capable of containing more electricity than its natural quantity. Consequently, when it is removed from the lower plate, and the whole of its electricity equally diffused through it, it will appear to have a redundance, and therefore will give a spark to any body communicating with the earth. Being then replaced upon the electric, and touched by any body communicating with the earth, it will be again affected as before, and give a spark on being raised; and this process may be continued at pleasure, the electrophorus supplying the place of any other electrical machine.

If the conducting plate of the electrophorus be applied to a piece of dry wood, marble, or any other substance through which electricity can pass but very slowly, or if the insulated conducting plate be covered with a piece of thin silk, which will make some resistance to the passage of electricity, and it be then applied to another plate communicating with the earth; and if, in either of these cases, a body with a large surface possessed of a weak electricity be applied to the conducting plate, the weak electricity not being able to overcome the obstruction presented to it, so as to be communicated to the other plate, will affect it with the contrary electricity, and this reacting on the first plate, will condense its electricity on that part of the plate to which it is contiguous; in consequence of which its capacity of receiving electricity will be increased; so that on the separation of the two plates, that electricity which was before condensed, being equally diffused through the whole plate, will have a greater intensity than it had before, attracting light bodies, or even giving a spark, when the body from which it received its electricity was incapable of it. For though it contained a great quantity of electricity, it was diffused through so large a space that its intensity was very small. This instrument is therefore called a *condenser of electricity*.

If an insulated plate of metal possessing the smallest degree of electricity be presented very near to another plate communicating with the earth, it will affect this plate with the opposite electricity; and this being, in the same manner, applied to a third plate, will put it into the same state with the first. If then these two plates be joined, and the first plate be presented to either of them, its own electricity being attracted by that of the plate presented, that of the other will be drawn into it, so that its quantity will be doubled. The same process being repeated, will again double the electricity of this plate, till, from being quite insensible to the most exquisite electrometer, it will become very conspicuous, or even give sparks. This instrument is therefore called a *doubler of electricity*, of excellent use in ascertaining the quality of atmospherical electricity when ever so small. If this instrument be so constructed that these three plates can be successively presented to one another by the revolution of one of them on an axis, it is called the *revolving doubler*; and in this form it is most convenient for use.

> [1] One gentleman of the Roman Catholic persuasion, and
> several of the Church of England, are now in the College.